How to Finance
Energy Managment Projects

How to Finance Energy Managment Projects

Solving the "Lack of Capital Problem"

Eric A. Woodroof, Ph.D., C.E.M., CRM
Albert Thumann, P.E., C.E.M.

Routledge
Taylor & Francis Group

LONDON AND NEW YORK

Published 2020 by River Publishers
River Publishers
Alsbjergvej 10, 9260 Gistrup, Denmark
www.riverpublishers.com

Distributed exclusively by Routledge
4 Park Square, Milton Park, Abingdon, Oxon OX14 4RN
605 Third Avenue, New York, NY 10017, USA

First issued in paperback 2023

Library of Congress Cataloging-in-Publication Data

Thumann, Albert.
How to finance energy managment projects : solving the "lack of capital problem" / Albert Thumann, Eric A. Woodroof. -- 1st ed.
    p. cm.
Includes bibliographical references and index.
ISBN 0-88173-701-1 (alk. paper) -- ISBN 978-8-7702-2304-1 (electronic : alk. paper) -- ISBN 978-1-4665-7153-2 (taylor & francis distribution : alk. paper) 1. Energy conservation--Finance. 2. Project management. I. Woodroof, Eric A. II. Title.

HD9502.A2T5196 2012
658.2'6--dc23

2012025797

How to Finance Energy Managment Projects / Eric A. Woodroof, Albert Thumann.
First published by Fairmont Press in 2013.

Routledge is an imprint of the Taylor & Francis Group, an informa business

Publisher's Note
The publisher has gone to great lengths to ensure the quality of this reprint but points out that some imperfections in the original copies may be apparent.

ISBN 978-87-7022-917-3 (pbk)
ISBN 978-1-4665-7153-2 (hbk)
ISBN 978-8-7702-2304-1 (online)
ISBN 978-1-0031-5173-9 (ebook master)

While every effort is made to provide dependable information, the publisher, authors, and editors cannot be held responsible for any errors or omissions.

for Bella...

Table of Contents

Foreword

As Executive Director of the Association of Energy Engineers (AEE), I have witnessed major changes in project financing. In 1977 when AEE was started, performance contracting was not widely accepted—energy project financing relied on "shared savings" but lacked the measurement and verification component. With the passage of the National Energy Policy Act of 1992, the federal government spearheaded the acceptance of "Energy Service Performance Contracting." Today the energy manager can avoid many of the financing pitfalls of the past and utilize proven performance contracting methodologies.

How to Finance Energy Management Projects provides indispensable guidance on how to overcome barriers, how to choose the right financing options, and how to understand a performance contract agreement. By using this reference, the reader will gain the tools necessary to implement energy efficiency and clean energy projects.

Albert Thumann, PE, CEM
Executive Director
Association of Energy Engineers

Contributors

CHAPTERS 1, 2, 8, 9, & 10

Eric A. Woodroof, Ph.D., is completely committed to helping businesses and organizations "go green," while improving profits. For more than 20 years, he has helped over 400 organizations and governments improve profits with energy/environmental solutions, generating over $100 million in savings.

Beyond his contributions as a consultant/project developer, he has taught over 100 seminars to help educate thousands of engineers worldwide on the best practices of energy and carbon management. These courses are endorsed by regions as diverse as Hong Kong, South Africa, and Chile. One of his goals is to "educate an army" of professionals who will have an even greater positive impact. The positive impact of this training is difficult to measure, but is sure to be an even greater contribution.

Dr. Woodroof served as the 2011 president of the Association of Energy Engineers, which is present in over 80 countries. He is the chairman of the board for the Certified Carbon Reduction Manager (CRM) program and he has been a board member of the Certified Energy Manager (CEM) Program since 1999. Dr. Woodroof has advised clients such as the U.S. government, airports, utilities, cities, universities and foreign governments. Private clients include numerous utilities, IBM, Frito-Lay, Pepsi, Ford, GM, Verizon, Hertz, Visteon, PriceSmart, Battelle and Lockheed Martin. His work has appeared in hundreds of articles and he has also delivered keynote speeches for clients on 6 continents.

Dr. Woodroof is a strategic advisor, corporate trainer and keynote speaker. He is the founder of ProfitableGreenSolutions.com and his direct line is 888-563-7221.

CHAPTER 3

Neil Zobler, President of Catalyst Financial Group, Inc., has been designing energy finance programs and arranged project-specific financing for demand side management (DSM) and renewable energy projects since 1985. Catalyst, a specialist in energy and water conservation projects, has arranged financings for over $1 billion. Neil's clients include U.S. EPA ENERGY STAR, the Inter-American Development Bank, over 20 electric and gas utilities (including Con Edison Co. of NY, PG&E, TVA), engineering companies and vendors, and hundreds of individual companies and organizations. He speaks regularly for organizations including the Government Finance Officers Association, the Association of School Business Officials, National Association of State Energy Officers, Association of Government Leasing & Finance, and the Council of Great City Schools and is on the task force of The American College & University Presidents Climate Commitment/ Clinton Climate Initiative program. He has been published widely in finance and energy periodicals. Neil is fluent in Spanish and helped design financing programs for energy projects in Mexico, Peru and El Salvador. Neil has a BA in Finance from Long Island University (LIU) and has completed post-graduate studies in marketing at the Arthur T. Roth Graduate School at LIU. His email address is nzobler@catalyst-financial.com.

Caterina (Katy) Hatcher is the US EPA ENERGY STAR National Manager for the Public Sector. She works with education, government, water and wastewater utility partners to help them improve their energy performance through the use of ENERGY STAR tools and resources. Katy has been working for the US Environmental Protection Agency for about 11 years. She holds a degree from the University of Virginia 's School of Architecture in City Planning.

EPA offers ENERGY STAR to organizations as a straightforward way to adopt superior energy management and realize the cost savings and environmental benefits that can result. EPA's guidelines for energy management promote a strategy for superior energy management that starts with the top leadership, engages the appropriate employees throughout the organization, uses standardized measurement tools, and helps an organization prioritize and get the most from its efficiency investments.

EPA's ENERGY STAR Challenge is a national call-to-action to improve the energy efficiency of America 's commercial and industrial facilities by 10 percent or more. EPA estimates that if the energy efficiency of commercial and industrial buildings and plants improved by 10 percent, Americans would save about $20 billion and reduce greenhouse gas emissions equal to the emissions from about 30 million vehicles.

CHAPTER 4

Shirley J. Hansen, Ph.D., is CEO of Hansen Associates, Inc. Shirley is widely recognized nationally and internationally for her expertise in energy management and energy efficiency financing. She has consulted for energy service companies (ESCOs), governments, multi-lateral development banks, and potential customers. She has trained hundreds of professionals, including end users, bankers and ESCO personnel, to use this financing mechanism effectively. Active in the U.S. and abroad, she has conducted workshops, developed manuals, and offered consultation in 38 countries in energy performance contracting, marketing analysis, business plan development, M&V financing, and sustainability management.

She has chaired the board of directors of the International Performance Measurement and Verification Protocol, Inc., and the certification board for M&V professionals. She is currently a member of the advisory board of the Association of Energy Engineers, the *Energy & Environmental Management* magazine, and the journal *Strategic Planning for Energy and the Environment.* She is also active in the International Working Group for the International Energy Efficiency Financing Protocol. She has authored and co-authored several books, including *Performance Contracting: Expanding Horizons, Investment Grade Energy Auditing, ESCO Around the World: Lessons Learned in 49 Countries,* and *Sustainability Management Handbook.* Her newest book, *World ESCO Outlook* is being released in 2012.

Shirley received her doctorate from Michigan State University. Among the many honors she has received, she is most proud of her distinguished alumni award from the university and her induction into the AEE Hall of Fame. She can be reached at kionaintl@aol.com.

CHAPTER 5

Ryan Park is one of the most influential individuals in the downstream integration market of the solar electricity industry. He was one of the founding members of REC Solar Inc., which now installs more solar electricity systems every year than any other company in the US.

Ryan is currently responsible for developing the commercial sales team, structuring large solar projects, and establishing new strategic partnerships. Ryan graduated with honors from California Polytechnic (Cal Poly), San Luis Obispo and is committed to improving the world through renewable energy technologies, energy efficiency, and empowering people.

His email address is: rpark@recsolar.com

James Coombes is a principal at Reveille Advisors. He has developed and structured financing solutions for commercial, government and residential solar PV projects since 2008. He entered the solar industry with REC Solar, contributing to project finance, strategic planning, and business development efforts. At REC Solar, he specialized in advising commercial and government customers on designing solar PV projects to optimize utility cost savings. He advised REC Solar's national accounts on developing solar projects across multiple markets to maximize financial returns. Prior to entering the solar industry, James was an investment banker for 10 years with Salomon Brothers and First Boston, advising cable television companies in the US, Japan and Europe. James later co-founded a beverage distribution and marketing company. He graduated from the University of California at Berkeley.

His email is jc@reveilleadvisors.com.

CHAPTER 6

Anthony J. Buonicore, PE, is a past president and Fellow Member of the Air & Waste Management Association, a Diplomat in the American Academy of Environmental Engineers, a Qualified Environmental Professional and a licensed professional engineer. He is a member of the ASTM Property Environmental Due Diligence committee, former chairman of its ASTM Phase I Task Group, and currently chairs the ASTM Task Group that developed the U.S. standard for vapor intrusion screening for properties involved in real estate transactions. In addition, Mr. Buonicore is chairman of the ASTM Task Group responsible for developing the new Building Energy Performance Assessment and Disclosure Standard.

Mr. Buonicore has been a leader in the energy-environmental industry since the early 1970s, serving as General Chairman of the American Institute of Chemical Engineers' First National Conference on Energy and the Environment in 1973 and as founder and first chairman of the Air Pollution Control Association's Energy-Environmental Interactions Technical Committee in 1974. He pioneered the use of refuse-derived fuel pellets (a bio-fuel) mixed with coal in stoker-fired boilers and has written extensively on energy and environmental issues.

As a Managing Director of Buonicore Partners, LLC, Mr. Buonicore is responsible for management of the firm's commercial real estate holdings and all due diligence activities associated with property acquisition. He holds both a bachelor's and a master's degree in chemical engineering.

CHAPTER 7

Jim Thoma is Managing Director for Green Campus Partners, LLC. He can be contacted at Raritan Plaza I, 110 Fieldcrest Avenue, Edison, NJ 08837; (732) 917-2303; jim.thoma@greencampuspartners.com

CHAPTER 8

John R. Wingender, Jr., is a Professor of Finance at Creighton University and the Chairman of the Department of Economics and Finance and can be reached at jwings@creighton.edu

He has served as the Associate Dean of Undergraduate Programs at Creighton University for three years. He previously taught at Oklahoma State University for twelve years where he was the Ardmore Professor of Business Administration. In 2005 he was a Fellow at the Institute of International Integration Studies at Trinity College Dublin, Ireland. He has been a Visiting Faculty member for the Beijing International MBA at Peking University in China since 2004. He received his Ph.D. from the University of Nebraska in 1985. He teaches managerial finance, international financial management, investment analysis, and portfolio management. Dr. Wingender has published more than 50 articles in refereed business journals, such as the Journal of Financial and Quantitative Analysis, Journal of Banking and Finance, Management Science, Journal of Business and Economic Statistics, Journal of Financial Research, The Financial Review, and the Journal of Business Research. He is an Editor of the Quarterly Journal of Finance and Accounting and an Associate Editor for the Quarterly Journal of Economics and Finance. He has been the President and officer of the Midwest Finance Association and the Association of Business Simulation and Experiential Learning. His research interests include the distribution of stock returns, the use of financial derivatives, counter cyclical hiring, and the decertification of unions. Dr. Wingender is currently the Treasurer of the Creighton Federal Credit Union Board of Directors. He serves on several nonprofit boards. At Creighton University he is the Faculty Advisor for the Financial Management Association (FMA) Student Chapter and the Phi Kappa Psi Fraternity. He has consulted for many businesses such as Ernst and Young, Ely Lily, Coca Cola, Dell Computers, and the World

Bank, most recently conducting a Financial Analysis training program for Pfizer, China. Dr. Wingender has two sons.

CHAPTER 9

Wayne C. Turner, Ph.D., PE, CEM, is a Regents Professor Emeritus of Industrial Engineering and Management at Oklahoma State University. He is founder and Director of OSU's Industrial Assessment Center and has conducted or supervised well over 1000 energy audits for industrial and commercial facilities. Dr. Turner has broad experience in energy management and has authored five textbooks and numerous articles in professional magazines and journals. He has won many teaching and professional awards and is listed in numerous *Who's Who* publications. He has served as past president of the Association of Energy Engineers (AEE) and is in AEE's Hall of Fame. He is Editor-In-Chief of AEE's journals *Energy Engineering* and *Strategic Planning for Energy and The Environment*. An avid fly fisherman, he is willing to fly fish anywhere, anytime. His email address is wayne.turner@okstate.edu

Warren M. Heffington, Ph.D., PE, CEM, is an associate professor emeritus of mechanical engineering at Texas A&M University and has directed an Industrial Assessment Center for 25 years. He has personally directed 250 industrial assessments and supervised the review of over 300 energy audit reports for commercial and institutional buildings. Research interests have included industrial energy use as well as the energy audit process. He teaches seminars on engineering ethics and professionalism and co-teaches the five-day Energy Management Fundamentals course offered by the Association of Energy Engineers.

Barney Capehart, Ph.D., CEM, is a Professor Emeritus of Industrial and Systems Engineering at the University of Florida, Gainesville. He has broad experience in the commercial/industrial sector, having served as Director of the University of Florida Industrial Assessment Center 1990-1999. He has personally conducted over 100 audits of industrial facilities and has assisted students in conducting audits of hundreds of office buildings, small businesses, government facilities, and other commercial facilities. He is the lead author for the *Guide to Energy Management* textbook, founding editor of the *Encyclopedia of Energy Engineering and Technology*, and co-author or editor of five other energy books. Dr. Capehart is the creator of AEE's Five Day Training Program for Energy Managers and has trained over 10,000 energy managers in that program.

CHAPTER 11

Millard Carr is SEA's Senior Vice President. As one of the most well-known and experienced experts in the federal energy programs, Mr. Carr can be consulted on virtually any energy-related subject at any level. He is recognized for an innovative and productive career of leadership for the Department of Defense, which operates the largest and most prolific corporate energy efficiency program in the world. He is a recovering Bureaucrat with over 32 years of experience in the federal government in utility engineering and procurement, mechanical engineering design and construction, life cycle cost-effective energy supply, and energy efficiency improvement. As the Head of the Utilities Management and Energy Programs Branch of the Naval Facilities Engineering Command (1973-1980), Mr. Carr was responsible for the development, coordination at all levels of Navy management and the implementation of the Navy's Energy Management Resource Plan to meet the requirements of Executive Order 12902 and the Energy Policy Act. Responsibilities included the review of component program plans and the development of policies to meet program goals and specific project oversight and management of the Navy's Energy Conservation Investment Program and Utilities Improvement Program, including approximately $70 million of projects per year. In this position, he initiated the use of alternatively financed energy projects for the development of renewable energy resources and shared energy savings contracts which evolved into Energy Savings Performance Contracts.

He retired from federal service in 1997 as the Director of Energy and Engineering, in the Office of the Secretary of Defense, overseeing energy programs totaling over $1.4 billion. Within the Office of the Secretary of Defense, Mr. Carr's responsibilities included providing guidance and leadership for design standards and construction criteria as well as utility procurement and energy efficiency improvement in the Department of Defense (DoD). As Director of Energy and Engineering, Mr. Carr was responsible for the development of DoD's policies and field-level implementation procedures. Mr. Carr also served as the DoD's representative on the Interagency Energy Management Working Group established by Congress, and was an active participant in the development of all National energy legislation including the Energy Policy Act of 1992 and Executive Orders 12759, 12902, and 13123 on federal energy efficiency. Mr. Carr has received numerous awards from federal and professional organizations and was inducted into AEE's "Energy Management Hall of Fame."

He is dedicated to helping private and public sector clients work together as partners to develop mutually beneficial energy management strategies, and successfully implement related projects. Experience includes: assistance to federal agencies and their private sector partners in developing energy program implementation strategies, and assistance in the development, training, and implementation of programs for federal energy and water efficiency, alternative financing of Energy Savings Performance Contracts and Utility Energy Services Contracts, and Renewable Energy Technologies applications.

APPENDIX A

David B. Pratt, Ph.D., PE, is an associate professor and the undergraduate program director in the School of Industrial Engineering and Management at Oklahoma State University. He holds B.S., M.S., and Ph.D. degrees in industrial engineering. Prior to joining academia, he held technical and managerial positions in the petroleum, aerospace, and pulp and paper industries for over 12 years. He has served on the industrial engineering faculty at his *alma mater*, Oklahoma State University, for over 16 years. His research, teaching, and consulting interests include production planning and control, economic analysis, and manufacturing systems design. He is a registered Professional Engineer, an APICS Certified Fellow in production and inventory management, and an ASQ Certified Quality Engineer. He is a member of IIE, NSPE, APICS, INFORMS, and ASQ.

Part I

Why & How

Chapter 1

If You Read Anything…
Do This First

The "Why"

Dear Reader,

When I began working in the energy management and financing industry, "lack of capital" was the prevailing excuse why a good project (less than a 3-year payback) was not being approved/implemented. When I did my Ph.D. research on this topic, about 35% of projects were postponed/canceled due to "lack of upfront capital." Despite my best efforts to solve this problem throughout my career, today the postponed percentage is closer to 50%! Looking at this trend, I can say that I have not succeeded at this goal. Although the global economy does have an influence on the percent of projects that are being initiated/financed, I still believe strongly that we can do MUCH more in this area to get more projects implemented that save energy, improve economic competitiveness and enable environmental prosperity.

This book will help you understand and hopefully implement financing solutions to get more projects implemented. The book is organized into sections to help you find what you need quickly, and then "just do it." But before I provide an outline to the whole book, I want to mention a few key concepts that you may be able to use today. These are "quick thoughts" on "big picture" concepts that I think you should keep in mind on all projects.

Woody's Winning Way (Key Concepts for Success):
1.   Presentation Point: If your energy project has an internal rate of return (IRR) that is greater than your company's profit margin, then the energy project is the best place to invest. *This is often the case as many energy projects have IRRs > 25%.*

2. Presentation Point: If your energy project has a return that is greater than the finance rate (borrowing rate), then you can finance the project (zero upfront cost) and you will improve cash flow to your organization, with relatively little risk.

3. Presentation Point: "savings = waste." Any energy savings that you could be getting via a potential project is also an existing waste stream that (by doing nothing) continues to drain your operating cash and is essentially a penalty you pay every month. Most people will take quicker action to avoid a penalty than to receive an equivalently valued reward. *To read a whole article on this point (and other articles) click the "Resources" tab at www.ProfitableGreenSolutions.com... the articles and webinars there are free.*

4. Presentation Point: "The cost of delay is usually greater than the cost of financing."

5. FYI: Know the codes, standards and laws that are driving activity in your building sector/geographic region. Whether it's the federal government or a local energy efficiency requirement, these "rules" can keep your project moving forward, as well as motivate projects that you never imagined. *See Chapter 11 for more.*

6. FYI: Know where you can get free money for your projects. This can be tax credits/deductions, utility rebates, special energy financing rates, utility energy service contracts, etc. *See www.DSIREusa.org for a list by state. Also see EERE and FEMP websites (just Google those acronyms)... very useful info.*

7. Presentation Point: Your audience only has the attention span to solve one problem at a time... *Make your presentation the most exiting solution possible...* so they can't resist approving it.

I can't stress how important your work is... and how much I value your efforts to implement energy efficiency projects. In my opinion, there are very few endeavors in today's capitalistic world, where "the more you do, the better." Your progress in energy efficiency does all of the following:

• Reduces your organization's expenses (and improves cash flow);
• Reduces your country's dependence on energy sources;
• Reduces your country's dependence on raw materials;
• Reduces your environmental impact (less pollution);
• Improves your organization's "green image";

- Likely improves your building's value (if leasing or selling);
- Likely improves your organization's morale and productivity.

When you are presenting your project, remember the above, because your project is *not just an energy project*; it may positively impact the marketing, administrative, finance, legal, human resources, security, and productivity departments too. If I have learned anything from the hundreds of organizations I have analyzed, or the thousands of students I have taught, one common denominator of behavior is evident: *"Necessity is the mother of invention."* Basically, if your project is "needed" by more of those departments, more people will be in the mood to approve your ideas.

I will offer you a "trade": My friends call me "Woody"—it has been a nickname since high school. If you implement some energy management projects in your local area, then you can call me Woody too, as you will become my friend. I also encourage you to let me know about your success; maybe I will share it with some other people around the world and they may be inspired by you! This way, you are making an even bigger impact.

In any event, I hope you can use this book to make a big difference, and NEVER GIVE UP on what is important to you!

Sincerely,

Eric Woodroof, Ph.D., CEM, CRM
Eric@ProfitableGreenSolutions.com

P.S.

OK, now on to the organization of the book. The chapters are organized into three parts, with some useful appendices at the end of the book.

In Part I, we cover the need for financing as well as the basic concepts.

In Part II, we present some chapters that were written by field experts. They cover some practical applications of financing such as performance contracts, power purchase agreements and other items like PACE financing. All of these "vehicles" of progress are innovative financial applications with proven success records. I want to mention Chapter 3 as it also covers some very useful tools that exist within the Energy Star® program. Chapter 7 shows you a financier's perspective and this can be quite helpful in planning the deal and avoiding mistakes.

Part III contains some very popular articles that have helped many engineers get more projects implemented. These articles also have more information that can be used to present projects and get them approved. I think these chapters are important because "<u>financing" is a "logical" solution; however, people purchase most items based on emotion</u>. There is more that can be said about this, but Part III will give you some ideas on how to leverage "non-logical" (and "logical") benefits.

Appendix A is basically about the "time value of money" fundamentals. Some may call this topic "engineering economy" or "Interest Rates 101." If you are brand new to financing, you may want to review this material. *Also, there is a recorded webinar (under the "Resources" tab at ProfitableGreenSolutions.com) that may help if you like to learn outside of a book.*

Appendix B is very short and has links to documents that are long and updated frequently. Thus, to save some paper required to make this book, use the links to get the M&V information that you need.

Appendix C is for those who may not know what an energy audit should look like. For many financiers, this information can be helpful in understanding what is a common deliverable from an engineer who is supposed to be doing a "Level II Audit." If you are an engineer, you will probably be bored reading this appendix, which is why it is an appendix… (Just trying to have some fun here!)

Appendix D is a sample of a project development agreement. These are used by ESCOs to engage the customer in the early development phases of a project. It is basically a vehicle for the ESCO to invest time, intellectual capital and resources into a potential client without fear that the intellectual capital will be wasted. Also called a "feasibility study," it is essentially a qualification tool. I feel it is helpful to understand this type of document because the development costs may also need to be financed so the client does not have to spend money out of pocket.

Appendix E is a sample of a performance contract. It is not perfect, or all-inclusive, but an example of a typical contract.

Appendix F provides additional explanation to clarify the sample performance contract.

Chapter 2

A Simple Introduction to Financing Energy Management Projects

Eric A. Woodroof, Ph.D., CEM, CRM

INTRODUCTION

Financing can be a key success factor for projects. This chapter's purpose is to help facility managers understand and apply the financial arrangements available to them. Hopefully, this approach will increase the implementation rate of good energy management projects, which would have otherwise been cancelled or postponed due to lack of funds.

Most facility managers agree that energy management projects (EMPs) are good investments. Generally, EMPs reduce operational costs, have a low risk/reward ratio, usually improve productivity, and even have been shown to improve a firm's stock price.[1] Despite these benefits, many cost-effective EMPs are not implemented due to financial constraints. Several studies show that first-cost and capital constraints are the main reasons why cost-effective EMPs are not implemented. Often, the facility manager does not have enough cash to allocate funding or cannot get budget approval to cover initial costs. Financial arrangements can mitigate a facility's funding constraints,[2] allowing additional energy savings to be reaped.

Alternative finance arrangements can overcome the initial cost obstacle, allowing firms to implement more EMPs. However, many facility managers are either unaware or have difficulty understanding the variety of financial arrangements available to them. Most facility managers use simple payback analyses to evaluate projects, which do not reveal the added value of cash flows that occur after the simple payback period has been reached.[3] Sometimes facility managers do not implement an EMP because financial terminology and contractual details intimidate them.[4]

7

Numerous papers and government programs (ex. EPA's Energy-Star tools) have been developed to show facility managers how to use quantitative (economic) analysis to evaluate financial arrangements.[3,4,5] *Quantitative analysis includes computing the simple payback, net present value (NPV), internal rate of return (IRR), and life-cycle cost of a project with or without financing.* Although these books and programs show how to evaluate the economic aspects of projects, they do not incorporate qualitative factors like strategic company objectives, which can impact the financial arrangement selection. Without incorporating a facility manager's qualitative objectives, it is hard to select an arrangement that meets all of the facility's needs. A recent paper showed that qualitative objectives can be at least as important as quantitative objectives.[8]

This chapter hopes to provide some valuable information that can be used to overcome the previously mentioned issues. The chapter is divided into several sections to accomplish three objectives. These sections will *introduce the basic financial arrangements* via a simple example and *define financial terminology.* For the purpose of this book, "financing" refers to different ways to fund a project without upfront capital (in most cases, like a home mortgage).* "Performance Contracting" generally means that the contractor must deliver some type of performance (usually energy savings, etc.) over a certain period after equipment is installed. Performance contracts can utilize a variety of financing mechanisms (loans, bonds, leases, etc.) and are generally more popular in the government/institutional types of facilities as banks are more willing to lend to facilities that are less vulnerable to the economy. Each arrangement is explained in greater detail while applied to a case study. The remaining sections show *how to match financial arrangements to different projects and facilities.* For those who need a more detailed description of rate of return analysis and basic financial evaluations, refer to Appendix A.

FINANCIAL ARRANGEMENTS: A SIMPLE EXAMPLE

Consider a small company, "PizzaCo," that makes frozen pizzas and distributes them regionally. PizzaCo uses an old delivery truck that

---

*It is important to note that when financing government or non-taxable entities, tax deductions would generally not apply.

breaks down frequently and is inefficient. Assume the old truck has no salvage value and is fully depreciated. PizzaCo's management would like to obtain a new and more efficient truck to reduce expenses and improve reliability. However, they do not have the cash on hand to purchase the truck. Thus, they consider their financing options.

Purchase the Truck with a Loan or Bond

Just like most car purchases, PizzaCo borrows money from a lender (a bank) and agrees to a monthly re-payment plan. Figure 2-1 shows PizzaCo's annual cash flows for a loan. The solid arrows represent the financing cash flows between PizzaCo and the bank. Each year, PizzaCo makes payments on the principal, plus interest based on the unpaid balance, until the balance owed is zero. The payments are the negative cash flows. Thus, at time zero when PizzaCo borrows the money, it receives a large sum of money from the bank, which is a positive cash flow that will be used to purchase the truck.

The *dashed* arrows represent the truck purchase as well as savings cash flows. Thus, at time zero, PizzaCo purchases the truck (a negative cash flow) with the money from the bank. Due to the new truck's greater efficiency, PizzaCo's annual expenses are reduced, which is a savings. The annual savings are the positive cash flows. The remaining cash flow diagrams in this chapter utilize the same format.

PizzaCo could also purchase the truck by selling a bond. This arrangement is similar to a loan, except investors (not a bank) give PizzaCo a large sum of money (called the bond's "par value"). Periodically, PizzaCo would pay the investors only the interest accumulated.

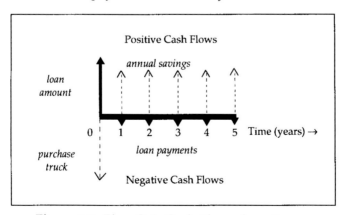

Figure 2-1. PizzaCo's Cash Flows for a Loan.

As Figure 2-2 shows, when the bond reaches maturity, PizzaCo returns the par value to the investors. The equipment purchase and savings cash flows are the same as with the loan.

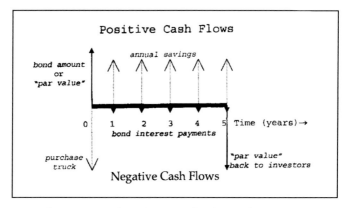

Figure 2-2. PizzaCo's Cash Flows for a Bond.

Sell Stock to Purchase the Truck

In this arrangement, PizzaCo sells its stock to raise money to purchase the truck. In return, PizzaCo is expected to pay dividends back to shareholders. Selling stock has a similar cash flow pattern as a bond, with a few subtle differences. Instead of interest payments to bondholders, PizzaCo would pay dividends to shareholders until some future date when PizzaCo could buy the stock back. However, these dividend payments are not mandatory, and if PizzaCo is experiencing financial strain, it is not required to distribute dividends. On the other hand, if PizzaCo's profits increase, this wealth will be shared with the new stockholders, because they now own a part of the company.

Rent the Truck

Just like renting a car, PizzaCo could rent a truck for an annual fee. This would be equivalent to a "true lease" or "operating lease." The rental company (lessor) owns and maintains the truck for PizzaCo (the lessee). PizzaCo pays the rental fees (lease payments), which are considered tax-deductible business expenses.

Figure 2-3 shows that the lease payments (solid arrows) start as soon the equipment is leased (year zero) to account for lease payments paid in advance. Lease payments "in arrears" (starting at the end of the first year) could also be arranged. However, the leasing company

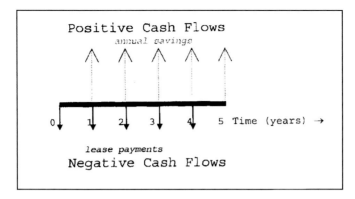

Figure 2-3. PizzaCo's Cash Flows for a True Lease.

may require a security deposit as collateral. Notice that the savings cash flows are essentially the same as the previous arrangements, except there is no equipment purchase, which is a large negative cash flow at year zero.

In a true lease, the contract period should be shorter than the equipment's useful life. The lease is cancelable because the truck can be leased easily to someone else. At the end of the lease, PizzaCo can either return the truck or renew the lease. In a separate transaction, PizzaCo could also negotiate to buy the truck at the fair market value.

If PizzaCo wanted to secure the option to buy the truck (for a bargain price) at the end of the lease, then they would use a capital lease. A capital lease can be structured like an installment loan, however ownership is not transferred until the end of the lease. The lessor retains ownership as security in case the lessee (PizzaCo) defaults on payments. Because the entire cost of the truck is eventually paid, the lease payments are larger than the payments in a true lease, (assuming similar lease periods). Figure 2-4 shows the cash flows for a capital lease with advance payments and a bargain purchase option at the end of year five.

There are some additional scenarios for lease arrangements. A "vendor-financed" agreement is when the lessor (or lender) is the equipment manufacturer. Alternatively, a third party could serve as a financing source. With "third-party financing," a finance company would purchase a new truck and lease it to PizzaCo. In either case, there are two primary ways to repay the lessor:

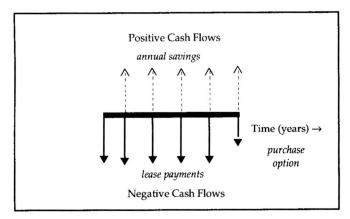

Figure 2-4. PizzaCo's Cash Flows for a Capital Lease.

1.   With a "fixed payment plan," where payments are due whether or not the new truck actually saves money.

2.   With a "flexible payment plan," where the savings from the new truck are shared with the third party until the truck's purchase cost is recouped with interest.

Subcontract Pizza Delivery to a Third Party

Since PizzaCo's primary business is not delivery, it could subcontract that responsibility to another company that would guarantee results (savings and/or performance). Let's say that a delivery service company would provide a truck and deliver the pizzas at a reduced cost. Each month, PizzaCo would pay the delivery service company a fee. However, this fee is guaranteed to be less than what PizzaCo would have spent on delivery. Thus, PizzaCo would obtain savings without investing any money or risk in a new truck. This arrangement is analogous to a performance contract. A performance contract can take many forms; however, the "performance" aspect is usually backed by a guarantee on operational performance from the contractor. In some performance contracts, the host can own the equipment and the guarantee assures that the operational benefits are greater than the finance payments. Alternatively, some performance contracts can be viewed as "outsourcing," where the contractor owns the equipment and provides a "service" to the host.

This arrangement is very similar to a third-party lease. However, with a performance contract, the contractor assumes most of the risk, and the contractor is also responsible for ensuring that the delivery fee is less than what PizzaCo would have spent. For the PizzaCo example, the arrangement would be designed under the conditions below:

- The delivery company is responsible for all operations related to delivering the pizzas. It also purchases, owns, and maintains the truck.

- The monthly fee is related to the number of pizzas delivered. This is the performance aspect of the contract; if PizzaCo doesn't sell many pizzas, the fee is reduced. *A minimum amount of pizzas may be required by the delivery company (performance contractor) to cover costs.* Thus, the delivery company assumes these risks:

  1. PizzaCo will remain solvent, and

  2. PizzaCo will sell enough pizzas to cover costs, and

  3. The new truck will operate as expected and will actually reduce expenses per pizza, and

  4. The external financial risk, such as inflation and interest rate changes, are acceptable.

- The delivery company is an expert in delivery; it has specially skilled personnel and uses efficient equipment. Thus, the delivery company can deliver the pizzas at a lower cost (even after adding a profit) than PizzaCo.

Figure 2-5 shows the net cash flows according to PizzaCo. Since the delivery company simply reduces PizzaCo's operational expenses,

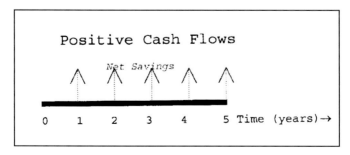

Figure 2-5. PizzaCo's Cash Flows for a Performance Contract.

there is only a net savings. There are no negative cash flows. Unlike the other arrangements, the delivery company's fee is a less expensive substitute for PizzaCo's in-house delivery expenses. With the other arrangements, PizzaCo had to pay a specific financing cost (loan, bond or lease payments, or dividends) associated with the truck, whether or not the truck actually saved money. In addition, PizzaCo would have to spend time maintaining the truck, which would detract from its core focus—making pizzas. With a performance contract, the delivery company is paid from the operational savings it generates. Because the savings are greater than the fee, there is a net savings. Often, the contractor guarantees the savings.

Supplementary note: Combinations of the basic finance arrangements are possible within a performance contract. For example, a guaranteed arrangement can be structured within a performance contract. Also, performance contracts are often designed so that the facility owner (PizzaCo) would own the asset at the end of the contract.

FINANCIAL ARRANGEMENTS: DETAILS AND TERMINOLOGY

To explain the basic financial arrangements in more detail, each one is applied to an energy management-related case study. To understand the economics behind each arrangement, some finance terminology is presented below.

Finance Terminology

Equipment can be purchased with cash on-hand (officially labeled "retained earnings"), a loan, a bond, a capital lease, or by selling stock. Alternatively, equipment can be utilized with a true lease or with a performance contract.

Note that with performance contracting, the building owner might not be paying for the equipment itself but the benefits provided by the equipment. *In the Simple Example, the benefit was the pizza delivery. PizzaCo was not concerned with what type of truck was used.*

The decision to purchase or utilize equipment is partly dependent on the company's strategic focus. If a company wants to delegate some or all of the responsibility of managing a project, it should use a true lease, or a performance contact.[9] However, if the company wants to be intricately involved with the EMP, purchasing and self-managing the equipment could yield the greatest profits. When the building owner purchases equipment, he/she usually maintains the equipment and lists it as an asset on the balance sheet so it can be depreciated.

Financing for purchases has two categories:

1. *Debt Financing,* which is borrowing money from someone else or another firm (using loans, bonds and capital leases).

2. *Equity Financing,* which is using money from your company or your stockholders (using retained earnings, or issuing common stock).

In all cases, the borrower will pay an interest charge to borrow money. The interest rate is called the "cost of capital." The cost of capital is essentially dependent on three factors: (1) the borrower's credit rating, (2) project risk and (3) external risk. External risk can include energy price volatility and industry-specific economic performance, as well as global economic conditions and trends. The cost of capital (or "cost of borrowing") influences the return on investment. If the cost of capital increases, then the return on investment decreases.

The "minimum attractive rate of return" (MARR) is a company's "hurdle rate" for projects. *Because many organizations have numerous projects competing for funding, the MARR can be much higher than interest earned from a bank or other risk-free investment.* Only projects with a return on investment greater than the MARR should be accepted. The MARR is also used as the discount rate to determine the "net present value" (NPV).

Explanation of Figures and Tables

Throughout this chapter's case study, figures are presented to illustrate the transactions of each arrangement. Tables are also presented to show how to perform the economic analyses of the different arrangements. The NPV is calculated for each arrangement.

It is important to note that the NPV of a particular arrangement can change significantly if the cost of capital, MARR, equipment residual value, or project life is adjusted. Thus, the examples within this chapter are provided only to illustrate how to perform the analyses. The cash flows and interest rates are estimates, which can vary from project to project. To keep the calculations simple, end-of-year cash flows are used throughout this chapter.

Within the tables, the following abbreviations and equations are used:

$$EOY = \text{End of Year}$$
$$\text{Savings} = \text{Pre-tax Cash Flow}$$
$$\text{Depr.} = \text{Depreciation}$$
$$\text{Taxable Income} = \text{Savings} - \text{Depreciation} - \text{Interest Payment}$$
$$\text{Tax} = (\text{Taxable Income})^*(\text{Tax Rate})$$
$$\text{ATCF} = \text{After-tax Cash Flow} = \text{Savings} - \text{Total Payments} - \text{Taxes}$$

Table 2-1 shows the basic equations that are used to calculate the values under each column heading within the economic analysis tables.

Regarding depreciation, the "modified accelerated cost recovery system" (MACRS) is used in the economic analyses. This system indicates the percent depreciation claimable year-by-year, after the equipment is purchased. Table 2-2 shows the MACRS percentages for seven-year property. *For example, after the first year, an owner could depreciate 14.29% of an equipment's value. The equipment's "book value" equals the remaining unrecovered depreciation. Thus, after the first year, the book value would be 100%-14.29%, which equals 85.71% of the original value. If the owner sells the property before it has been fully depreciated, he/she can claim the book value as a tax-deduction.**

APPLYING FINANCIAL ARRANGEMENTS:
A CASE STUDY

Suppose PizzaCo (*the "host" facility*) needs a new chilled water system for a specific process in its manufacturing plant. The installed

---

To be precise, the IRS uses a "half-year convention" for equipment that is sold before it has been completely depreciated. In the tax year that the equipment is sold, (say year "x") the owner claims only Ω of the MACRS depreciation percent for that year. (This is because the owner has only used the equipment for a fraction of the final year.) Then on a separate line entry, (in the year "x"), the remaining unclaimed depreciation is claimed as "book value." The x* year is presented as a separate line item to show the book value treatment, however x* entries occur in the same tax year as "x."

Table 2-1. Table of Sample Equations used in Economic Analyses.

A	B	C	D	E	F	G	H	I	J
			Payments						
EOY	*Savings*	*Depreciation*	*Principal*	*Interest*	*Total*	*Principal Outstanding*	*Taxable Income*	*Tax*	*ATCF*
n									
n+1		= (MACRS %)* (Purchase Price)			=(D) +(E)	=(G at year n) -(D at year n+1)	=(B)-(C)-(E)	=(H)*(tax rate)	=(B)-(F)-(I)
n+2									

Table 2-2. MACRS Depreciation Percentages.

EOY	MACRS Depreciation Percentages for 7-Year Property
0	0
1	14.29%
2	24.49%
3	17.49%
4	12.49%
5	8.93%
6	8.92%
7	8.93%
8	4.46%

cost of the new system is $2.5 million. The expected equipment life is 15 years, however the process will only be needed for 5 years, after which the chilled water system will be sold at an estimated market value of $1,200,000 (book value at year five = $669,375). The chilled water system should save PizzaCo about $1 million/year in energy savings. PizzaCo's tax rate is 34%. The equipment's annual maintenance and insurance cost is $50,000. PizzaCo's MARR is 18%. Since at the end of year 5, PizzaCo expects to sell the asset for an amount greater than its book value, the additional revenues are called a "capital gain" (equals the market value – book value) and are taxed. If PizzaCo sells the asset for less than its book value, PizzaCo incurs a "capital loss."

PizzaCo does not have $2.5 million to pay for the new system, thus it considers its finance options. PizzaCo is a small company with an average credit rating, which means that it will pay a higher cost of capital than a larger company with an excellent credit rating. (As with any borrowing arrangement, if investors believe that an investment is risky, they will demand a higher interest rate.)

Purchase Equipment with Retained Earnings (Cash)

If PizzaCo did have enough retained earnings (cash on-hand) available, it could purchase the equipment without external financing. Although external finance expenses would be zero, the benefit of tax deductions from interest expenses is also zero. Also, any cash used to purchase the equipment would carry an "opportunity cost," because that cash could have been used to earn a return somewhere else. This opportunity cost rate is usually set equal to the MARR. In other words, the company lost the opportunity to invest the cash and gain at least the MARR from another investment.

Of all the arrangements described in this chapter, purchasing equipment with retained earnings is probably the simplest to understand. For this reason, it will serve as a brief example and introduction to the economic analysis tables that are used throughout this chapter.

Application to the Case Study

Figure 2-6 illustrates the resource flows between the parties. In this arrangement, PizzaCo purchases the chilled water system directly from the equipment manufacturer.

Once the equipment is installed, PizzaCo recovers the full $1 million/year in savings for the entire five years, but it must spend $50,000/year on maintenance and insurance. At the end of the five-year project, PizzaCo expects to sell the equipment for its market value of $1,200,000. Assume MARR is 18% and the equipment is classified as 7-year property for MACRS depreciation. Table 2-3 shows the economic analysis for purchasing the equipment with retained earnings.

Reading Table 2-3 from left to right, and top to bottom, at EOY 0, the single payment is entered into the table. Each year thereafter, the savings as well as the depreciation (which equals the equipment purchase price multiplied by the appropriate MACRS % for each year)

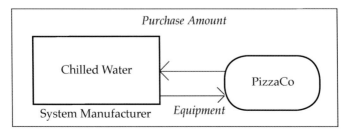

Figure 2-6. Resource Flows for Using Retained Earnings

Table 2-3. Economic Analysis for Using Retained Earnings (Cash).

EOY	Savings	Depr.	Payments Principal	Interest	Total	Principal Outstanding	Taxable Income	Tax	ATCF
0					2,500,000				-2,500,000
1	950,000	357,250					592,750	201,535	748,465
2	950,000	612,250					337,750	114,835	835,165
3	950,000	437,250					512,750	174,335	775,665
4	950,000	312,250					637,750	216,835	733,165
5	950,000	111,625					838,375	285,048	664,953
5*	1,200,000	669,375					530,625	180,413	1.019.588
		2,500,000							

Net Present Value at 18%: $320,675

Notes: Loan Amount: 0

          Loan Finance Rate: 0%

          MARR   18%

          Tax Rate   34%

MACRS Depreciation for 7-Year Property, with half-year convention at EOY 5

Accounting Book Value at end of year 5:   669,375

Estimated Market Value at end of year 5:   1,200,000

EOY 5* illustrates the Equipment Sale and *Book* Value

          Taxable Income: =(Market Value - Book Value)

                           =(1,200,000 - 669,375) = $530,625

are entered into the table. Year by year, the taxable income = savings – depreciation. The taxable income is then taxed at 34% to obtain the tax for each year. The after-tax cash flow = savings – tax for each year.

At EOY 5, the equipment is sold before the entire value was depreciated. EOY 5* shows how the equipment sale and book value are claimed. In summary, the NPV of all the ATCFs (after tax cash flow) would be $320,675.

Loans

Loans have been the traditional financial arrangement for many types of equipment purchases. A bank's willingness to loan depends on the borrower's financial health, experience in energy management, and number of years in business. Obtaining a bank loan can be difficult if the loan officer is unfamiliar with EMPs. Loan officers and financiers may not understand energy-related terminology (demand charges, kVAR, etc.). In addition, facility managers may not be comfortable with the financier's language. Thus, to save time, a bank that can understand EMPs should be chosen.

Most banks will require a down payment and collateral to secure a loan. However, securing assets can be difficult with EMPs, because the equipment often becomes part of the real estate of the plant. *For example, it would be very difficult for a bank to repossess lighting fixtures from a retrofit.* In these scenarios, lenders may be willing to secure other assets as collateral.

Application to the Case Study

Figure 2-7 illustrates the resource flows between the parties. In this arrangement, PizzaCo purchases the chilled water system with a loan from a bank. PizzaCo makes equal payments (principal + interest) to

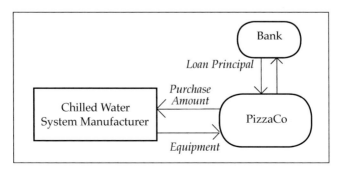

2-7. Resource Flow Diagram for a Loan.

the bank for five years to retire the debt. Due to PizzaCo's small size, credibility, and inexperience in managing chilled water systems, PizzaCo is likely to pay a relatively high cost of capital. For example, let's assume 15%.

PizzaCo recovers the full $1 million/year in savings for the entire five years, but it must spend $50,000/year on maintenance and insurance. At the end of the five-year project, PizzaCo expects to sell the equipment for its market value of $1,200,000. Tables 2-4 and 2-5 show the economic analysis for loans with a zero down payment and a 20% down payment, respectively. Assume that the bank reduces the interest rate to 14% for the loan with the 20% down payment. Since the asset is listed on PizzaCo's balance sheet, PizzaCo can use depreciation benefits to reduce the after-tax cost. In addition, all loan interest expenses are tax-deductible.

Bonds

Bonds are very similar to loans; a sum of money is borrowed and repaid with interest over a period of time. The primary difference is that with a bond, the issuer (PizzaCo) periodically pays the investors only the interest earned. This periodic payment is called the "coupon interest payment." *For example, a $1,000 bond with a 10% coupon will pay $100 per year. When the bond matures, the issuer returns the face value ($1,000) to the investors.*

Bonds are issued by corporations and government entities. Government bonds generate tax-free income for investors, thus these bonds can be issued at lower rates than corporate bonds. This benefit provides government facilities an economic advantage to use bonds to finance projects.

Application to the Case Study

Although PizzaCo (a private company) would not be able to obtain the low rates of a government bond, they could issue bonds with coupon interest rates competitive with the loan interest rate of 15%.

In this arrangement, PizzaCo receives the investors' cash (bond par value) and purchases the equipment. PizzaCo uses part of the energy savings to pay the coupon interest payments to the investors. When the bond matures, PizzaCo must then return the par value to the investors. (See Figure 2-8.)

As with a loan, PizzaCo owns, maintains and depreciates the

Table 2-4. Economic Analysis for a Loan with No Down Payment.

EOY	Savings	Depr.	Payments			Principal Outstanding	Taxable Income	Tax	ATCF
			Principal	Interest	Total				
0						2,500,000			
1	950,000	357,250	370,789	375,000	745,789	2,129,211	217,750	74,035	130,176
2	950,000	612,250	426,407	319,382	745,789	1,702,804	18,368	6,245	197,966
3	950,000	437,250	490,368	255,421	745,789	1,212,435	257,329	87,492	116,719
4	950,000	312,250	563,924	181,865	745,789	648,511	455,885	155,001	49,210
5	950,000	111,625	648,511	97,277	745,789	0	741,098	251,973	-47,761
5*	1,200,000	669,375					530,625	180,413	1,019,588
		2,500,000							

Net Present Value at 18%: $757,121

Notes: Loan Amount: 2,500,000 (used to purchase equipment at year 0)
Loan Finance Rate: 15%   MARR   18%
                                           Tax Rate   34%

MACRS Depreciation for 7-Year Property, with half-year convention at EOY 5
Accounting Book Value at end of year 5: 669,375
Estimated Market Value at end of year 5: 1,200,000
EOY 5* illustrates the Equipment Sale and Book Value
        Taxable Income: =(Market Value - Book Value)
                              =(1,200,000 - 669,375) = $530,625

Table 2-5. Economic Analysis for a Loan with a 20% Down Payment,

EOY	Savings	Depr.	Payments Principal	Payments Interest	Payments Total	Principal Outstanding	Taxable Income	Tax	ATCF
0					500,000	2,000,000			–500,000
1	950,000	357,250	302,567	280,000	582,567	1,697,433	312,750	106,335	261,098
2	950,000	612,250	344,926	237,641	582,567	1,352,507	100,109	34,037	333,396
3	950,000	437,250	393,216	189,351	582,567	959,291	323,399	109,956	257,477
4	950,000	312,250	448,266	134,301	582,567	511,024	503,449	171,173	196,260
5	950,000	111,625	511,024	71,543	582,567	0	766,832	260,723	106,710
5*	1,200,000	669,375					530,625	180,413	1,019,588
		2,500,000							

Net Present Value at 18%: $710,962

Notes: Loan Amount: 2,000,000 (used to purchase equipment at year 0)
        Loan Finance Rate: 14%   MARR   18%
                                                            500,000   Tax Rate   34%

MACRS Depreciation for 7-Year Property, with half-year convention at EOY 5
Accounting Book Value at end of year 5:   669,375
Estimated Market Value at end of year 5:   1,200,000
EOY 5* illustrates the Equipment Sale and *Book Value*
        Taxable Income: =(Market Value - Book Value)
                            =(1,200,000 - 669,375) = $530,625

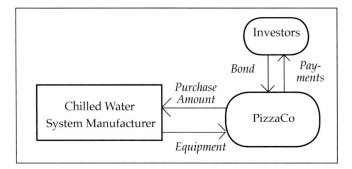

Figure 2-8. Resource Flow Diagram for a Bond.

equipment throughout the project's life. All coupon interest payments are tax-deductible. At the end of the five-year project, PizzaCo expects to sell the equipment for its market value of $1,200,000. Table 2-6 shows the economic analysis of this finance arrangement.

Selling Stock

    Although less popular, selling company stock is an equity financing option which can raise capital for projects. For the host, selling stock offers a flexible repayment schedule, because dividend payments to shareholders aren't absolutely mandatory. Selling stock is also often used to help a company attain its desired capital structure. However, selling new shares of stock dilutes the power of existing shares and may send an inaccurate "signal" to investors about the company's financial strength. If the company is selling stock, investors may think that it is desperate for cash and in a poor financial condition. Under this belief, the company's stock price could decrease. However, recent research indicates that when a firm announces an EMP, investors react favorably.[10] On average, stock prices were shown to increase abnormally by 21.33%.

    By definition, the cost of capital (rate) for selling stock is:

cost of capital$_{\text{selling stock}}$ = D/P
where   *D = annual dividend payment*
           P = company stock price

    However, in most cases, the after-tax cost of capital for selling stock is higher than the after-tax cost of debt financing (using loans, bonds and capital leases). This is because interest expenses (on debt) are tax deductible, but dividend payments to shareholders are not.

Table 2-6. Economic Analysis for a Bond.

EOY	Savings	Depr.	Payments Principal	Payments Interest	Total	Principal Outstanding	Taxable Income	Tax	ATCF
0						2,500,000			
1	950,000	357,250		375,000	375,000	2,500,000	217,750	74,035	500,965
2	950,000	612,250		375,000	375,000	2,500,000	-37,250	-12,665	587,665
3	950,000	437,250		375,000	375,000	2,500,000	137,750	46,835	528,165
4	950,000	312,250		375,000	375,000	2,500,000	262,750	89,335	485,665
5	950,000	111,625	2,500,000	375,000	2,875,000	0	463,375	157,548	-2,082,548
5*	1,200,000	669,375					530,625	180,413	1,019,588
		2,500,000							

Net Present Value at 18%:   953,927

Notes: Loan Amount: 2,500,000 (used to purchase equipment at year 0)
        Loan Finance Rate: 0%   MARR   18%
                                                       Tax Rate   34%

MACRS Depreciation for 7-Year Property, with half-year convention at EOY 5
Accounting Book Value at end of year 5:   669,375
Estimated Market Value at end of year 5:   1,200,000
EOY 5* illustrates the Equipment Sale and *Book Value*
            Taxable Income: =(Market Value - Book Value)
                                 =(1,200,000 - 669,375) = $530,625

In addition to tax considerations, there are other reasons why the cost of debt financing is less than the financing cost of selling stock. Lenders and bond buyers (creditors) will accept a lower rate of return because they are in a less risky position due to the reasons below.

- Creditors have a contract to receive money at a certain time and future value. (Stockholders have no such guarantee with dividends.)

- Creditors have first claim on earnings. (Interest is paid before shareholder dividends are allocated.)

- Creditors usually have secured assets as collateral and have first claim on assets in the event of bankruptcy.

Despite the high cost of capital, selling stock does have some advantages. This arrangement does not bind the host to a rigid payment plan (like debt financing agreements), because dividend payments are not mandatory. The host has control over when it will pay dividends. Thus, when selling stock, the host receives greater payment flexibility, but at a higher cost of capital.

Application to the Case Study

As Figure 2-9 shows, the financial arrangement is very similar to a bond. At year zero the firm receives $2.5 million, except the funds come from the sale of stock. Instead of coupon interest payments, the firm distributes dividends. At the end of year five, PizzaCo repurchases the stock. Alternatively, PizzaCo could capitalize the dividend payments, which means setting aside enough money so the dividends could be paid with the interest generated.

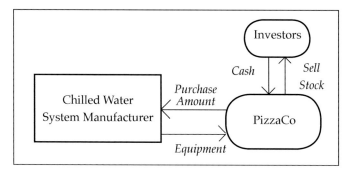

Figure 2-9. Resource Flow Diagram for Selling Stock.

Table 2-7 shows the economic analysis for issuing stock at a 16% cost of equity capital, then repurchasing the stock at the end of year five. (For consistency of comparison to the other arrangements, the stock price does not change during the contract.) Like a loan or bond, PizzaCo owns and maintains the asset. Thus, the annual savings are only $950,000. PizzaCo pays annual dividends worth $400,000. At the end of year 5, PizzaCo expects to sell the asset for $1,200,000.

Note that Table 2-7 is slightly different from the other tables in this chapter:

Taxable Income = Savings – Depreciation, and

ATCF = Savings – Stock Repurchases - Dividends - Tax

Leases

Firms generally own assets, however it is the use of these assets that is important, not the ownership. Leasing is another way of obtaining the use of assets. There are numerous types of leasing arrangements, ranging from basic rental agreements to extended payment plans for purchases. Table 2-8 lists some additional reasons why leasing can be an attractive arrangement for the lessee.

Basically, there are two types of leases: the "true lease" (a.k.a. "operating" or "guideline lease") and the "capital lease." One of the primary differences between a true lease and a capital lease is the tax treatment. In a true lease, the lessor owns the equipment and receives the depreciation benefits. However, the lessee can claim the entire lease payment as a tax-deductible business expense. In a capital lease, the lessee (PizzaCo) owns and depreciates the equipment. However, only the interest portion of the lease payment is tax-deductible. In general, a

Table 2-8. Good Reasons to Lease.

Financial Reasons
- With some leases, the entire lease payment is tax-deductible.
- Some leases allow "off-balance sheet" financing, preserving credit lines

Risk Sharing
- Leasing is good for short-term asset use, and reduces the risk of getting stuck with obsolete equipment
- Leasing offers less risk and responsibility

Table 2-7. Economic Analysis of Selling Stock.

EOY	Savings	Depr.	Stock Transactions Sale of Stock	Repurchase	Dividend Payments	Taxable Income	Tax	ATCF
0			$2,500,000 from Stock Sale is used to purchase equipment, thur ATCF = 0					
1	950,000	357,250			400,000	592,750	201,535	348,465
2	950,000	612,250			400,000	337,750	114,835	435,165
3	950,000	437,250			400,000	512,750	174,335	375,665
4	950,000	312,250			400,000	637,750	216,835	333,165
5	950,000	111,625		2,500,000	400,000	838,375	285,048	-2,235,048
5*	1,200,000	669,375				530,625	180,413	1,019,588
		2,500,000						

Net Present Value at 18%: 477,033

Notes: Value of Stock Sold (which is repurchased after year five 2,500,000 (used to purchase equipment at year zero)

Cost of Capital = Annual Dividend Rate: 16%   MARR =   18%

                                                       Tax Rate =   34%

MACRS Depreciation for 7-Year Property, with half-year convention at EOY 5

Accounting Book Value at end of year 5:   669,375

Estimated Market Value at end of year 5:   1,200,000

EOY 5* illustrates the Equipment Sale and Book Value

         Taxable Income: = (Market Value - Book Value)

                                 = (1,200,000 - 669,375) = $530,625

true lease is effective for a short-term project, where the company does not plan to use the equipment when the project ends. A capital lease is effective for long-term equipment.

The True Lease

Figure 2-10 illustrates the legal differences between a true lease and a capital lease. A true lease (or operating lease) is strictly a rental agreement. The word "strictly" is appropriate because the Internal Revenue Service will only recognize a true lease if it satisfies the following criteria:

1.   The lease period must be less than 80% of the equipment's life.

2.   The equipment's estimated residual value must be 20% of its value at the beginning of the lease.

3.   There is no "bargain purchase option."

4.   There is no planned transfer of ownership.

5.   The equipment must not be custom-made nor useful only in a particular facility.

Application to the Case Study

It is unlikely that PizzaCo could find a lessor that would be willing to lease a sophisticated chilled water system and, after five years, move the system to another facility. Thus, obtaining a true lease would be unlikely. Nevertheless, Figure 2-11 shows the basic relationship between the lessor and lessee in a true lease. A third-party leasing company could also be involved by purchasing the equipment and leasing to PizzaCo. Such a resource flow diagram is shown for the capital lease.

Table 2-9 shows the economic analysis for a true lease. Notice that the lessor pays the maintenance and insurance costs, so PizzaCo saves the full $1 million per year. PizzaCo can deduct the entire lease payment of $400,000 as a business expense. However, PizzaCo does not obtain ownership, so it can't depreciate the asset.

The Capital Lease

The capital lease has a much broader definition than a true lease. A capital lease fulfills *any one* of the following criteria:

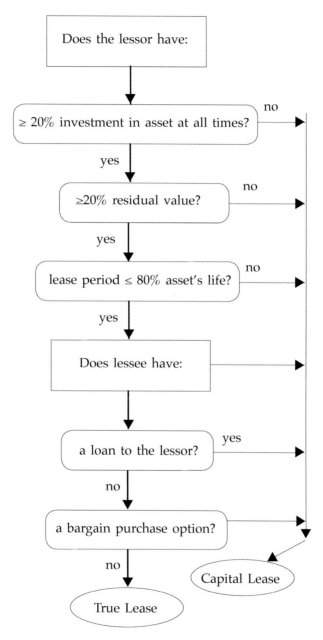

Figure 2-10. Classification for a True Lease.

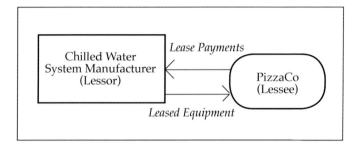

Figure 2-11. Resource Flow Diagram for a True Lease.

1. The lease term ≥75% of the equipment's life.
2. The present value of the lease payments ≥ 90% of the initial value of the equipment.
3. The lease transfers ownership.
4. The lease contains a "bargain purchase option" that is negotiated at the inception of the lease.

Most capital leases are basically extended payment plans, except ownership is usually not transferred until the end of the contract. This arrangement is common for large EMPs, because the equipment (such as a chilled water system) is usually difficult to reuse at another facility. With this arrangement, the lessee eventually pays for the entire asset (plus interest). In most capital leases, the lessee pays the maintenance and insurance costs.

The capital lease has some interesting tax implications, because the lessee must list the asset on its balance sheet from the beginning of the contract. Thus, like a loan, the lessee gets to depreciate the asset, and only the interest portion of the lease payment is tax deductible.

Application to the Case Study

Figure 2-12 shows the basic third-party financing relationship between the equipment manufacturer, lessor, and lessee in a capital lease. The finance company (lessor) is shown as a third party, although it also could be a division of the equipment manufacturer. Because the finance company (with excellent credit) is involved, a lower cost of capital (12%) is possible, due to reduced risk of payment default.

Like an installment loan, PizzaCo's lease payments cover the

Table 2-9. Economic Analysis for a True Lease

EOY	Savings	Depr.	Payments Principal	Payments Interest	Total	Principal Outstanding	Taxable Income	Tax	ATCF
0							-400,000		-400,000
1	1,000,000				400,000		600,000	204,000	396,000
2	1,000,000				400,000		600,000	204,000	396,000
3	1,000,000				400,000		600,000	204,000	396,000
4	1,000,000				400,000		600,000	204,000	396,000
5	1,000,000						1,000,000	340,000	660,000

Net Present Value at 18%: $953,757

Notes: Annual Lease Payment: 400,000
MARR = 18%
Tax Rate 34%

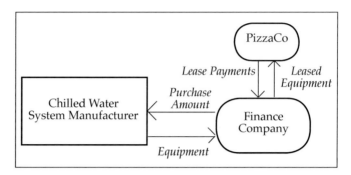

Figure 2-12. Resource Flow Diagram for a Capital Lease.

entire equipment cost. However, the lease payments are made in advance. Because PizzaCo is considered the owner, it pays the $50,000 annual maintenance expenses, which reduces the annual savings to $950,000. PizzaCo receives the benefits of depreciation and tax-deductible interest payments. To be consistent with the analyses of the other arrangements, PizzaCo would sell the equipment at the end of the lease for its market value. Table 2-10 shows the economic analysis for a capital lease.

Performance Contracting

Performance contracting is a unique arrangement that allows the building owner to make necessary improvements while investing very little money upfront. The contractor usually assumes responsibility for purchasing and installing the equipment, as well as maintenance throughout the contract. But the unique aspect of performance contracting is that the contractor is also responsible for the performance of the installed equipment. In some cases, only after the new equipment actually reduces expenses does the contractor get paid. Energy service companies (ESCOs) typically serve as consultants and contractors within this line of business. Usually, the performance contracting process begins when the ESCO conducts an energy audit of the host's facilities to determine if the savings are large enough to warrant further effort. If so, then the ESCO and host often enter into a Project Development Agreement (PDA). In the PDA phase, the ESCO commits greater resources and literally develops a detailed proposal (ESCO takes measurements to develop detailed estimates of costs and guaranteed savings). The host then signs the formal Perfor-

Table 2-10. Economic Analysis for a Capital Lease.

EOY	Savings	Depr.	Payments in Advance			Principal Outstanding	Taxable Income	Tax	ATCF
			Principal	Interest	Total				
0			619,218	0	619,218	1,880,782		-619,218	
1	950,000	357,250	393,524	225,694	619,218	1,487,258	367,056	124,799	205,983
2	950,000	612,250	440,747	178,471	619,218	1,046,511	159,279	54,155	276,627
3	950,000	437,250	493,637	125,581	619,218	552,874	387,169	131,637	199,145
4	950,000	312,250	552,874	66,345	619,218	0	571,405	194,278	136,503
5	950,000	111,625					838,375	285,048	664,953
5*	1,200,000	669,375					530,625	180,413	1,019,588
		2,500,000				Net Present Value at 18%:			$681,953

Notes:

Total Lease Amount: 2,500,000

However, Since the payments are in advance, the first payment is analogous to a Down-Payment

Thus the actual amount borrowed is only = $500,000 - 619,218 = 1,880,782

Lease Finance Rate: 12%   MARR   18%

Tax Rate   34%

MACRS Depreciation for 7-Year Property, with half-year convention at EOY 5

Accounting Book Value at end of year 5:   669,375

Estimated Market Value at end of year 5:   1,200,000

EOY 5* illustrates the Equipment Sale and Book Value

Taxable Income: =(Market Value - Book Value)

=(1,200,000 - 669,375) = $530,625

mance Contract, which might last several years. At the same time, the host (or possibly ESCO) secures financing such that the cash flow is positive for the project. *Note that during the entire development process (which sometimes takes years), the host may choose to cancel the agreement and pay a fee to the ESCO to compensate for their intellectual property, opportunity costs and/or investment into the host. All of these "penalty costs" should be agreed upon in advance of signing a PDA.*

Again, unlike most loans, leases and other fixed payment arrangements, the ESCO is paid based on the performance of the equipment. In other words, if the finished product doesn't save energy or operational costs, the host doesn't pay. This aspect removes the incentive to "cut corners" on construction or other phases of the project, as with bid/ spec contracting. In fact, often there is an incentive to exceed savings estimates. For this reason, performance contracting usually entails a more "facility-wide" scope of work (to find extra energy savings) than loans or leases on particular pieces of equipment.

With a facility-wide scope, many improvements can occur at the same time. For example, lighting and air conditioning systems can be upgraded at the same time. In addition, the indoor air quality can be improved. With a comprehensive facility management approach, a "domino effect" on cost reduction is possible. For example, if facility improvements create a safer and higher quality environment for workers, productivity could increase. As a result of decreased employee absenteeism, the workman's compensation cost could also be reduced. These are additional benefits to the facility.

Depending on the host's capability to manage the risks (equipment performance, financing, etc.) the host will delegate some of these responsibilities to the ESCO. In general, the amount of risk assigned to the ESCO is directly related to the percent savings that must be shared with the ESCO.

For facilities that are not in a good position to manage the risks of an energy project, performance contracting may be the only economically feasible implementation method. *For example, the US federal government uses performance contracting to upgrade facilities when budgets are being dramatically cut. In essence, they "sell" some of their future energy savings to an ESCO, in return for receiving new equipment and efficiency benefits.*

In general, performance contracting may be the best option for facilities that:

- are severely constrained by their cash flows;

- have a high cost of capital;

- don't have sufficient resources, such as a lack of in-house energy management expertise or an inadequate maintenance capacity*;

- are seeking to reduce in-house responsibilities and focus more on their core business objectives; or

- are attempting a complex project with uncertain reliability, or if the host is not fully capable of managing the project. *For example, a lighting retrofit has a high probability of producing the expected cash flows, whereas a completely new process does not have the same "time-tested" reliability. If the in-house energy management team cannot manage this risk, performance contracting may be an attractive alternative.*

Performance contracting does have some drawbacks. In addition to sharing the savings with an ESCO, the tax benefits of depreciation and other economic benefits must be negotiated. Whenever large contracts are involved, there is reason for concern. One study found that 11% of customers who were considering EMPs felt that dealing with an ESCO was too confusing or complicated.[11] Another reference claims, "With complex contracts, there may be more options and more room for error." Therefore, it is critical to choose an ESCO with a good reputation and experience within the types of facilities that are involved.

There are a few common types of contracts. The ESCO will usually offer the following options:

- guaranteed fixed dollar savings;
- guaranteed fixed energy (MMBtu) savings;
- a percent of energy savings; or
- a combination of the above.

Obviously, facility managers would prefer the options with "guaranteed savings." However this extra security (and risk to the ESCO) usually costs more. The primary difference between the two guaranteed

---

*Maintenance capacity represents the ability of the maintenance personnel to maintain the new system. It has been shown that systems fail and are replaced when maintenance concerns are not incorporated into the planning process. See Woodroof, E. (1997) "Lighting Retrofits: Don't Forget About Maintenance," Energy Engineering, 94(1) pp. 59-68.

options is that guaranteed fixed dollar savings contracts ensure dollar savings, even if energy prices fall. *For example, if energy prices drop and the equipment does not save as much money as predicted, the ESCO must pay (out of its own pocket) the contracted savings to the host.*

Percent energy savings contracts are agreements that basically share energy savings between the host and the ESCO. The more energy saved, the higher the revenues to both parties. However, the host has less predictable savings and must also periodically negotiate with the ESCO to determine "who saved what" when sharing savings. Although there are numerous hybrid contracts available that combine the positive aspects of the above options.

Periodically, the host and ESCO need to measure and verify (M&V) that the savings are "on target." Usually a "baseline" of energy consumption is established that represents the host's facility performance before the retrofit. Once the baseline is established and the energy efficient equipment (or other change to facility) is implemented, then the new energy consumption is measured and verified (by a third party) to ensure that the savings are actually occurring.

There are a variety of M&V strategies and guidelines/protocols that establish a "rule book" on how to deal with variances, which almost always will occur. *For example, if the host is a growing university and has doubled enrollment 5 years after the energy efficient equipment is installed, then there might not be any "savings," but there may be "avoided costs" from an "adjusted baseline"(which represents what the university would have spent without the energy efficient equipment).*

Sometimes if a performance contract is not "on target," the ESCO and Host can reconcile the differences in a variety of ways, such as extending the terms, savings "buy-outs" or the ESCO offering additional services (or cash) to make up a shortfall of savings (or avoided costs). The M&V guidelines have prescriptive and mostly fair ways to deal with these issues. M&V guidelines have been developed from lots of experiences, however when an ESCO has "boilerplate" language regarding the M&V methods, the host should review that language carefully to be sure that it is fair to both parties.

Appendix B has a link to the most popular M&V Guidelines.

Application to the Case Study

PizzaCo would enter into a hybrid contract: *percent energy savings/ guaranteed arrangement.* The ESCO would purchase, install and operate a highly efficient chilled water system. The ESCO would guarantee that PizzaCo would save the $1,000,000 per year, but PizzaCo would pay the ESCO 80% of the savings. In this way, PizzaCo would not need to invest any money and would simply collect the net savings of $200,000 each year. To avoid periodic negotiations associated with shared savings agreements, the contract could be worded such that the ESCO will provide guaranteed energy savings worth $200,000 each year.

With this arrangement, there are no depreciation, interest payments or tax-benefits for PizzaCo. However, PizzaCo receives a positive cash flow with no investment and little risk. At the end of the contract, the ESCO removes the equipment. At the end of most performance contracts, the host usually acquires or purchases the equipment for fair market value; however, for this case study, the equipment was removed to make a consistent comparison with the other financial arrangements.

Figure 2-13 illustrates the transactions between the parties. These arrangements can take a variety of forms, but as long as savings is greater than finance payment, then performance contracting can work. Table 2-11 presents the economic analysis for performance contracting.

Note that Table 2-11 is slightly different from the other tables in this chapter: Taxable Income = Savings – Depreciation – ESCO Payments.

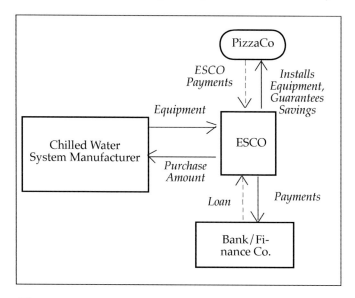

Figure 2-13. Transactions for a Performance Contract.

Table 2-11. Economic Analysis of a Performance Contract.

EOY	Savings	Depr.	ESCO Payments		Principal Outstanding	Taxable Income	Tax	ATCF
			Total					
0								
1	1,000,000		800,000			200,000	68,000	132,000
2	1,000,000		800,000			200,000	68,000	132,000
3	1,000,000		800,000			200,000	68,000	132,000
4	1,000,000		800,000			200,000	68,000	132,000
5	1,000,000		800,000			200,000	68,000	132,000

Net Present Value at 18%: $412,787

Notes: ESCO purchases/operates equipment. Host pays ESCO 80% of the savings = $800,000.
         The contract could also be designed so that PizzaCo can buy the equipment at the end of year 5.

Additional Options

Combinations of the basic financial arrangements can be created to enhance the value of a project. A sample of the possible combinations are described below.

• In some performance vontracts, the host can own the equipment and the guarantee assures that the operational benefits are greater than the finance payments. Alternatively, some performance contracts can be viewed as "outsourcing," where the contractor owns the equipment and provides a "service" to the host.

• Third-party financiers often cooperate with performance contracting firms to implement EMPs.

• Utility rebates and government programs may provide additional benefits for particular projects.

• Tax-exempt leases are available to government facilities.

• Insurance can be purchased to protect against risks relating to equipment performance, energy savings, etc.

• Some financial arrangements can be structured as non-recourse to the host. Thus, the ESCO or lessor would assume the risks of payment default. However, as mentioned before, profit sharing increases with risk sharing.

Attempting to identify the absolute best financial arrangement is a rewarding goal, unless it takes too long. As every minute passes, potential dollar savings are lost forever. When considering special grant funds, rebate programs, or other unique opportunities, it is important to consider the lost savings due to delay.

PROS & CONS OF EACH
FINANCIAL ARRANGEMENT

This section presents a brief summary of the pros and cons of each financial arrangement from the host's perspective.

Table 2-12. Pros and Cons of Financial Arrangements

LOAN

Pros

- Host keeps all savings.
- Depreciation and interest payments are tax-deductible.
- Host owns the equipment.
- The arrangement is good for long-term use of equipment.

Cons

- Host takes all the risk and must install and manage the project.

BOND

Same Pros/Cons as loan, plus:

Pro

- Good for government facilities, because they can offer a tax-free rate that is lower (but considered favorable by investors).

SELL STOCK

Same Pros/Cons as loan, plus:

Pros

- Selling stock could help the host achieve its target capital structure.

Cons

- Dividend payments (unlike interest payments) are not tax-deductible.
- Dilutes company control.

USE RETAINED EARNINGS

Same Pros/Cons as loan, plus:

Pros

- Host pays no external interest charges. However, retained earnings do carry an opportunity cost, because such funds could be invested somewhere at the MARR.

Cons

- Host loses tax-deductible benefits of interest charges

(Continued)

Table 2-12 *(Cont'd)*. Pros and Cons of Financial Arrangements

CAPITAL LEASE
Same Pros/Cons as loan, plus:
Pro
- Has greater flexibility in financing and possible lower cost of capital with third-party participation.

TRUE LEASE

Pros	*Cons*
• Allows use of equipment, without ownership risks.	• There is no ownership at the end of lease contract.
• Has reduced risk of poor performance, service, equipment obsolescence, etc.	• There are no depreciation tax benefits,
• Is good for short-term use of equipment.	
• Entire lease payment is tax-deductible.	

PERFORMANCE CONTRACT

Pros	*Cons*
• Allows use of equipment, with reduced installment/operational risks.	• Involves potentially binding contracts, legal expenses, and increased administrative costs.
• Reduced risk of poor performance or service, equipment obsolescence, etc.	• Host must share project savings.
• Allows host to focus on its core business objectives.	

Rules of Thumb

When investigating financing options, consider the following generalities:

Loans, bonds and other host-managed arrangements should be used when a customer has the resources (experience, financial support, time) to handle the risks. Performance contracting (ESCO assumes most of the risk) is usually best when a customer doesn't have the resources to properly manage the project. Remember that with any arrangement where the host delegates risk to another firm, the host must also share the savings.

Leases are the "middle ground" between owning and delegating risks, and they are very popular due to their tax benefits.

True leases tend to be preferred when:

- the equipment is needed on a short-term basis;

- the equipment has unusual service problems that cannot be handled by the host;

- technological advances cause equipment to become obsolete quickly; or

- depreciation benefits are not useful to the lessee.

Capital Leases are preferred when:

- the installation and removal of equipment is costly;

- the equipment is needed for a long time; or

- the equipment user desires to secure a "bargain purchase option."

INCORPORATING STRATEGIC ISSUES WHEN SELECTING FINANCIAL ARRANGEMENTS

Because strategic issues can be important when selecting financial arrangements, the facility manager should include them in the selection process. The following questions can help assess a facility manager's needs:

- Does the facility manager want to manage projects or outsource?
- Are net positive cash flows required?
- Will the equipment be needed for long-term needs?
- Is the facility government or private?
- If private, does the facility manager want the project's assets on or off the balance sheet?
- Will operations be changing?

From the research experience, a Strategic Issues Financing Decision Tree was developed to guide facility managers to the financial arrangement that is most likely optimal. Figure 2-14 illustrates the decision tree, which is by no means a rule, but it embodies some general observations from the industry.

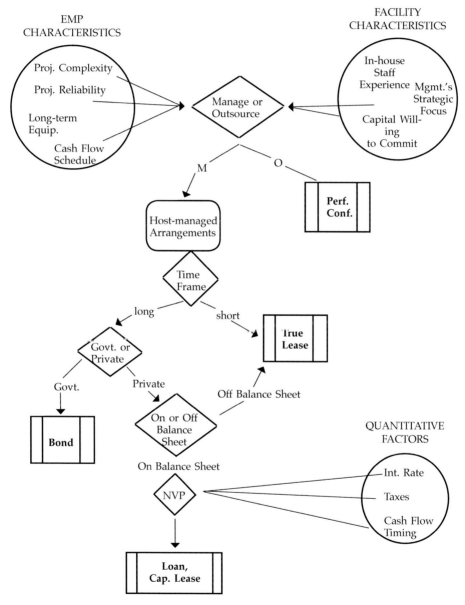

Figure 2-14. Strategic Issues Financing Decision Tree.

Table 2-13. Examples

FEDERAL GOVERNMENT

Facility Type: Location:
Military Installation   California

Project:
Performance Contract including over $40 million in energy efficiency equipment,
    which saves $7,000,000/year

Financial Arrangement: Capital Lease with a 20-year term

STATE AND LOCAL GOVERNMENT

Facility Type: Location:
Local Government-Airport   Tulsa, OK

Project:
Performance Contract including over $4 million in energy efficiency equipment,
    which saves $380,000/year

Financial Arrangement: Municipal Bond with a 20-year term

Facility Type: Location:
Local Government-Convention Center   Kansas City, MO

Project:
Performance Contract including over $8 million in energy efficiency equipment,
    which saves over $1 million/year

Financial Arrangement: Municipal Lease with a 10-year term

Education
Facility Type: Location:
University   New Orleans, LA

Project:
$8 million in energy-related upgrades

Financial Arrangement: Operating Lease (Synthetic) with a 20-year term

---

HEALTH CARE

Facility Type: Location:
Hospital   Memphis, TN

Project:
$15 million in energy-related upgrades

Financial Arrangement: Operating Lease (Synthetic) with a 20-year term

---

Working the tree from the top to bottom, the facility manager should assess the project and facility characteristics to decide whether it is strategic to manage the project or outsource. If outsourced, the "performance contract" would be the logical choice.* If the facility manager wants to manage the project, the next step (moving down the tree) is to evaluate whether the project's equipment will be needed for long or short-term purposes. If short-term, the "true lease" is logical. If it is a long-term project, in a government facility, the "bond" is likely to be the best option. If the facility is in the private sector, the facility manager should decide whether the project should be on or off the balance sheet. An off-balance sheet preference would lead back to the "true lease." If the facility manager wants the project's assets on the balance sheet, the Net Present Value (or other economic benefit indicator) can help determine which "host-managed" arrangement (loan, capital lease or cash) would be most lucrative.

Although the decision tree can be used as a guide, it is most important to use the financial arrangement that best meets the needs of the organization. The examples in Table 2-13 demonstrate that any organization can be creative with its financial arrangement selection. All of these examples are for Performance Contracting projects; however, similar financial arrangements can be structured without using a performance contract.

CHAPTER SUMMARY

It is clear that knowing the strategic needs of the facility manager is critical to selecting the best arrangement. There are practically an infinite number of financial alternatives to consider. This chapter has provided some information on the basic financial arrangements. Combining these arrangements to construct the best contract for your facility is only limited by your creativity.

---

*It should be noted that a performance contract could be structured using leases and bonds.

GLOSSARY

Capitalize

To convert a schedule of cash flows into a principal amount, called capitalized value, by dividing by a rate of interest. In other words, to set aside an amount large enough to generate (via interest) the desired cash flows forever.

Capital or Financial Lease

Lease that, under Statement 13 of the Financial Accounting Standards Board, must be reflected on a company's balance sheet as an asset and corresponding liability. Generally, this applies to leases where the lessee acquires essentially all of the economic benefits and risks of the leased property.

Depreciation

The amortization of fixed assets, such as plant and equipment, so as to allocate the cost over their depreciable life. Depreciation reduces taxable income but is not an actual cash flow.

Energy Service Company (ESCO)

Company that provides energy services (and possibly financial services) to an energy consumer.

Host

The building owner or facility that uses the equipment.

Lender

Individual or firm that extends money to a borrower with the expectation of being repaid, usually with interest. Lenders create debt in the form of loans or bonds. If the borrower is liquidated, the lender is paid off before stockholders receive distributions.

Lessee

The renter. The party that buys the right to use equipment by making lease payments to the lessor.

Lessor

The owner of the leased equipment.

Line of Credit

An informal agreement between a bank and a borrower indicating the maximum credit the bank will extend. A line of credit is popular because it allows numerous borrowing transactions to be approved without re-application paperwork.

Liquidity

Ability of a company to convert assets into cash or cash equivalents without significant loss. For example, investments in money market funds are much more liquid than investments in real estate.

Leveraged Lease

Lease that involves a lender in addition to the lessor and lessee. The lender, usually a bank or insurance company, puts up a percentage of the cash required to purchase the asset, usually more than half. The balance is put up by the lessor, who is both the equity participant and the borrower. With the cash the lessor acquires the asset, giving the lender (1) a mortgage on the asset and (2) an assignment of the lease and lease payments. The lessee then makes periodic payments to the lessor, who in turn pays the lender. As owner of the asset, the lessor is entitled to tax deductions for depreciation on the asset and interest on the loan.

MARR (Minimum Attractive Rate of Return)

MARR is the "hurdle rate" for projects within a company. MARR is used to determine the NPV. The annual after-tax cash flow is discounted at MARR (which represents the rate the company could have received with a different project).

Net Present Value (NPV)

As the saying goes, "A dollar received next year is not worth as much as a dollar today." The NPV converts the worth of that future dollar into what it is worth today. NPV converts future cash flows by using a given discount rate. For example, at 10%, $1,000 dollars received one year from now is worth only $909.09 dollars today. In other words, if you invested $909.09 dollars today at 10%, in one year it would be worth $1,000.

NPV is useful because you can convert future savings cash flows back to "time zero" (the present), and then compare to the cost of a project. If the NPV is positive, the investment is acceptable. In capital budgeting, the discount rate used is called the "hurdle rate" and is usually equal to the incremental cost of capital.

"Off-Balance Sheet" Financing

Typically refers to a True Lease, because the assets are not listed on the balance sheet. Because the liability is not on the balance sheet, the Host appears to be financially stronger. However, most large leases must be listed in the footnotes of financial statements, which reveals the "hidden" assets.

Par Value or Face Value

Equals the value of the bond at maturity. For example, a bond with a $1,000 dollar par value will pay $1,000 to the issuer at the maturity date.

Preferred Stock

A hybrid type of stock that pays dividends at a specified rate (like a bond), and has preference over common stock in the payment of dividends and liquidation of assets. However, if the firm is financially strained, it can avoid paying the preferred dividend as it would the common stock dividends. Preferred stock doesn't ordinarily carry voting rights.

Project Financing

A type of arrangement typically meaning that a Single Purpose Entity (SPE) is constructed. The SPE serves as a special bank account. All funds are sent to the SPE, from which all construction costs are paid. Then all savings cash flows are also distributed from the SPE. The SPE is essentially a mini-company, with the sole purpose of funding a project.

Secured Loan

Loan that pledges assets as collateral. Thus, in the event that the borrower defaults on payments, the lender has the legal right to seize the collateral and sell it to pay off the loan.

True Lease or Operating Lease or Tax-oriented Lease

Type of lease, normally involving equipment, whereby the contract is written for considerably less time than the equipment's life, and the lessor handles all maintenance and servicing; also called service lease. Operating leases are the opposite of capital leases, where the lessee acquires essentially all the economic benefits and risks of ownership. Common examples of equipment financed with operating leases are office copiers, computers, automobiles and trucks. Most operating leases are cancelable.

WACC (Weighted Average Cost of Capital)

The firm's average cost of capital, as a function of the proportion of different sources of capital: Equity, Debt, Preferred Stock, etc. *For example, a firm's target capital structure is:*

Capital Source	Weight (w_i)
Debt	30%
Common Equity	60%
Preferred Stock	10%

and the firm's costs of capital are:
    before tax cost of debt $= k_d \ = \ $ 10%
    cost of common equity $= k_s \ = \ $ 15%
    cost of preferred stock $= k_{ps} \ = \ $ 12%

Then the weighted average cost of capital will be:
    $\text{WACC} = w_d k_d (1\text{-}T) + w_s k_s + w_{ps} k_{ps}$

       where $w_i = weight\ of\ Capital\ Source_i$
               $T = tax\ rate = 34\%$
               After-tax cost of debt $= k_d(1\text{-}T)$

Thus,
    $\text{WACC} = (.3)(.1)(1\text{-}.34) + (.6)(.15) + (.1)(.12)$
    $\text{WACC} = 12.18\%$

References

1. Wingender, J. and Woodroof, E., (1997) "When Firms Publicize Energy Management Projects Their Stock Prices Go Up: How High?—As Much as 21.33% within 150 days of an Announcement," *Strategic Planning for Energy and the Environment*, Vol. 17(1), pp. 38-51.
2. Woodroof, E. and Turner, W. (1998), "Financial Arrangements for Energy Management Projects," *Energy Engineering* 95(3) pp. 23-71.
3. Sullivan, A. and Smith, K. (1993) "Investment Justification for U.S. Factory Automation Projects," *Journal of the Midwest Finance Association*, Vol. 22, p. 24.
4. Fretty, J. (1996), "Financing Energy-Efficient Upgraded Equipment," Proceedings of the 1996 International Energy and Environmental Congress, Chapter 10, Association of Energy Engineers.
5. Pennsylvania Energy Office, (1987) The Pennsylvania Life Cycle Costing Manual.
6. United States Environmental Protection Agency (1994). ProjectKalc, Green Lights Program, Washington DC.
7. Tellus Institute, (1996), P2/Finance version 3.0 for Microsoft Excel Version 5, Boston MA.

8. Woodroof, E. And Turner, W. (1999) "Best Ways to Finance Your Energy Management Projects," *Strategic Planning for Energy and the Environment*, Summer 1999, Vol. 19(1) pp. 65-79.

9. Cooke, G.W., and Bomeli, E.C., (1967), *Business Financial Management*, Houghton Mifflin Co., New York.

10. Wingender, J. and Woodroof, E., (1997) "When Firms Publicize Energy Management Projects: Their Stock Prices Go Up," *Strategic Planning for Energy and the Environment*, 17 (1) pp. 38-51.

11. Coates, D.F. and DelPonti, J.D. (1996), "Performance Contracting: a Financial Perspective" *Energy Business and Technology Sourcebook*, Proceedings of the 1996 World Energy Engineering Congress, Atlanta. p.539-543.

Part II

Tools & Techniques

Chapter 3

Choosing the Right Financing And How to Use ENERGY STAR® Tools

Neil Zobler
Catalyst Financial Group, Inc.
Katy Hatcher
U.S. Environmental Protection Agency,
ENERGY STAR Program

FINANCING EFFICIENCY PROJECTS TODAY
WITH FUTURE ENERGY SAVINGS

Administrators or managers often think they must postpone the implementation of energy efficiency upgrades because they do not have the funds in their current budgets. Other barriers may include the lack of time, personnel, or expertise. As you will see later in this chapter, most organizations have access to more financial resources than they may think, and resolving the financial barrier frequently provides solutions to all the other barriers.

Postponing energy efficiency upgrades for as little as one year can prove to be an expensive decision. The U.S. Environmental Protection Agency (EPA) estimates that as much as 30% of the energy consumed in buildings may be used unnecessarily or inefficiently. The money lost due to these inefficiencies in just one year frequently totals more than all the costs of financing energy upgrades over the course of the entire financing period!

Consider this business logic: The energy efficiency project you'd like to do is not in your current capital budget; but if the costs of financing the project are less than the operating budget dollars saved from reduced utility bills, why not finance the project? The benefits of doing the project sooner rather than later are numerous, starting with improved cash flow, better facilities, using the existing capital budget for other projects, helping make your facility "green," and more. Using this logic, financing energy efficiency projects is a good business decision. *Delaying the project is, however, a conscious decision to continue paying the utility companies for the energy waste* rather than investing these dollars in improving your facilities.

Energy efficiency projects are unlike most other projects. With properly structured financing, you may be able to implement energy efficiency projects without exceeding your existing operating or capital budgets. Thousands of companies that participate in ENERGY STAR know from experience that today's energy efficiency technologies and practices have saved them operating budget dollars. In fact, some of the more conservative money center lenders like Bank of America, JP Morgan Chase, and CitiBank have been providing funds for energy efficiency projects for over a decade. And implementing energy efficiency projects will have a positive impact on your organization's overall financial performance as well as the environment. So why wait? This chapter will help you understand how to leverage your savings opportunities, which financing vehicles to consider, and where to find the money.

OPERATING VERSUS CAPITAL BUDGETS

Before addressing different financing options and vehicles, let's review some "Accounting 101" fundamentals. Organizations make purchases by spending their own cash or by borrowing the needed funds. The impact on the balance sheet is either exchanging one asset (current asset: cash) for another (fixed asset: equipment) when spending cash, or adding to the assets *and* liabilities when incurring debt.

To argue the advantages of one financing option versus another, it is important to be conversant with the roles of the operating expense budget and the capital expense budget in your organization. *Capital expenses* are those that pay for long-term debt and fixed assets (such as buildings, furniture, school buses, etc.) and whose repayment typically extends **beyond** one operating period (one operating period usually being 12 months). In contrast, *operating expenses* are those general and operating expenses (such as salaries or supply bills) incurred **during** one operating period (again, typically 12 months)[1]. For example, repayment of a bond issue is considered a capital expense, whereas paying monthly utility bills is considered an operating expense.

The disadvantages associated with trying to use capital expense budget dollars for your energy efficiency projects are as follows: (1) current fiscal year capital dollars are usually already committed to other projects; (2) capital dollars are often scarce, so your efficiency projects are competing with other priorities; and (3) the approval process for requesting new capital dollars is time consuming, expensive, and often cumbersome.

When arranging financing for energy efficiency projects in the private sector, one of the most frequently asked questions is, "How do we keep this financing off our balance sheet?" and thereby not reflect the transaction on the company's financial statements as a liability or debt. The reasons for this request vary by organization and include: (a) treating the repayment of the obligation as an operating expense, thereby avoiding the entire capital budget process, and, (b) avoiding the need for compliance with restrictive covenants that are frequently imposed by existing lenders, which may be viewed as cumbersome to a point of interfering with the ongoing management of the company. Restrictive covenants start by requiring the borrower to periodically provide financial statements that enable the lender to track the performance of the company by calculating key financial ratios measuring liquidity (i.e., the current ratio, which is current assets versus current liabilities), leverage (debt-to-equity), and profitability margins. Covenants include maintaining financial ratios at agreed standards. If the ratios are not in compliance with these targets, the lender can call in all loans, creating serious cash flow problems for the borrower. Many of these financial ratios are improved by keeping debt off the balance sheet. Other typical covenants include limitations on issuing new debt, paying dividends to stockholders, and selling assets of the company.

While organizations in the public sector may not have to deal with the profitability and equity issues of the private sector, they do face their own challenges when incurring debt through the capital budget process, which is established by statute, constitution, or charter, and usually requires voter approval. Public-sector organizations may find that the political consequences of incurring new debt may be more of a deterrent than the financial ones, particularly when raising taxes is involved.

Treating repayment of the financing for energy efficiency projects as an operating expense can keep the financing "off the balance sheet." And, because the immediate benefit of installing energy efficiency projects is reducing the operating expense budget earmarked for paying the energy and water bills, off-balance-sheet treatment makes sense. Post-ENRON, however, having your auditors treat financing as "off balance sheet" is becoming increasingly difficult, especially in light of The Sarbanes-Oxley Act of 2002, which established new or enhanced standards for all U.S. public company boards, management, and public accounting firms.[2]

Nevertheless, several financing vehicles do allow financing payments for energy efficiency upgrades to be treated as operating expenses. (These include operating leases, power purchase agreements, and tax-exempt lease-purchase agreements; see the FINANCIAL INSTRUMENTS section for more information.) Regardless of the type of financing vehicle used, implementing energy efficiency projects is usually in the best financial interests of your organization, and is helping to protect the environment.

FINANCING TOOLS AND RESOURCES FROM ENERGY STAR

With the help of the ENERGY STAR program, American families and businesses have saved a combined total of nearly $230 billion on utility bills over the past 20 years. Their energy efficiency actions have kept more than 1.7 billion metric tons of greenhouse gases from being emitted into our skies. Today, ENERGY STAR is a household name. It's one of the most well known brands in the country, and more than 80 percent of Americans now recognize the blue ENERGY STAR label. EPA's ENERGY STAR program offers a proven strategy for superior energy management, with tools and resources to help each step of the way. Based on the successful practices of ENERGY STAR partners, EPA's Guidelines for Energy Management (available at www.energystar.gov/guidelines) illustrates how organizations can improve operations and maintenance strategies to reduce energy use and realize the cost savings that can be generated by financing energy efficiency projects. EPA has sponsored hundreds of presentations (in person and on the Internet) about ENERGY STAR tools, resources, and best practices for organizations struggling with the challenge of making their buildings more energy efficient. One of the most common statements from participants, especially those in the public sector, is "We don't have the money needed to do the facility upgrades; in fact, we don't even have enough money to pay for the energy audits needed to determine the size of the savings opportunity." This sentiment is simply not true because the needed funds are currently sitting in their utility operating budgets and being doled out every month to the local utilities. Organizations merely require a way to capture and redirect these "wasted energy" funds to pay for the energy efficiency projects, which will in turn create real savings. For some readers, this may seem to be "circular logic," or what may be called a "Catch 22."

Fortunately, ENERGY STAR has created a number of tools and resources that, when properly used, will allow you to "break the circle" and find a path toward the timely implementation of energy efficiency projects. This section focuses on the tools that tie directly into financing such projects. These include the Guidelines for Energy Management, EPA's ENERGY STAR Portfolio Manager™, Financial Value Calculator, Building Upgrade Value Calculator, and the Cash Flow Opportunity Calculator. All of these tools are in the public domain and available at www.energystar.gov. The main tools we discuss here are:

- Portfolio Manager (A Benchmarking Tool)
- The Building Upgrade Value Calculator (Financial Evaluation Tool)
- Decision Guide for State and Local Governments
- Cash Flow Opportunity Calculator (Cost of Delay Tool & Sensitivity Analysis Tool)

Portfolio Manager

Noted management expert Peter Drucker's famous maxim is, "If you don't measure it, you can't manage it." If your organization wants to use future energy savings to pay for the implementation of energy efficiency projects now, you must start by establishing the baseline of your current energy usage. EPA's Portfolio Manager can help you do that. It is an interactive energy management tool that allows you to track and assess energy and water consumption across your entire portfolio of buildings in a secure online environment. Portfolio Manager can help you identify under-performing buildings, set investment priorities, verify effectiveness of efficiency improvements, and receive EPA recognition for superior energy performance.

Any building manager or owner can efficiently track and manage resources through Portfolio Manager. The tool allows you to streamline your portfolio's energy and water data, as well as track key consumption, performance, and cost information portfolio-wide. For example, you can:

- Track multiple energy and water meters for each facility

- Customize meter names and key information

- Benchmark your facilities relative to their past performance

- View percentage improvement in weather-normalized source energy

- Monitor energy and water costs

- Share your building data with others inside or outside your organization

- Enter operating characteristics tailored to each space use category within a building

For many types of facilities, you can receive an ENERGY STAR energy performance score on a scale of 1-100 relative to similar buildings nationwide. Your building is *not* compared to the other buildings in Portfolio Manager to determine your ENERGY STAR score. Instead, statistically representative models are used to compare your building against similar buildings from a national survey conducted by the U.S. Department of Energy's (DOE's) Energy Information Administration known as the Commercial Building Energy Consumption Survey (CBECS). Conducted every four years, CBECS gathers data on building characteristics and energy use from thousands of buildings across the United States. Your building's peer group for comparison is the group of buildings in the CBECS survey that has similar construction and operating characteristics. An ENERGY STAR score of 50 indicates that the building, from an energy consumption standpoint, performs better than 50% of all similar buildings nationwide, while a score of 75 indicates that the building performs better than 75% of all similar buildings nationwide.

EPA's energy performance scoring system, based on source energy, accounts for the impact of weather variations, as well as changes in key physical and operating characteristics of each building.[3] Buildings that achieve a score of 75 or higher may qualify for ENERGY STAR certification.

Portfolio Manager provides a platform to track energy and water use trends compared against the costs of these resources. This is a valuable tool for understanding the relative costs associated with a given level of performance, helping you evaluate investment opportunities for a particular building, and identifying the best opportunities across your portfolio. It also allows you to track your properties' performance from year to year.

The built-in financial tool within Portfolio Manager helps you compare cost savings across buildings in your portfolio while calculating cost savings for a specific project. Being able to quickly and clearly obtain data showing cumulative investments in facility upgrades or annual energy costs eases the decision-making process for best practice management of your buildings nationwide.

From a lender's perspective, a facility with a low ENERGY STAR score is more likely to obtain larger energy savings (having more room for improvement) than a facility with a high score. This becomes important if the energy savings are considered a primary "source of repayment" when financing energy upgrades. Portfolio Manager is also an important tracking mechanism that helps ensure that the facilities are being properly maintained and the energy savings are continuing to accrue. As lenders perform due diligence on energy efficiency projects, they will become more aware of the value of Portfolio Manager.

Building Upgrade Value Calculator

The Building Upgrade Value Calculator is built on a Microsoft EXCEL™ platform and is a product of the partnership between EPA's ENERGY STAR program, Building Owners and Managers Association (BOMA) International, and the BOMA Foundation. The Building Upgrade Value Calculator estimates the financial impact of proposed investments in energy efficiency on office properties. The calculations are based on data input by the user, representing scenarios and conditions present at their properties. Required inputs are limited to general characteristics of the building, plus information on the proposed investments in energy efficiency upgrades. The calculator's analysis includes the following information:

- Net investment
- Reduction in operating expense
- Energy savings
- Return on investment (ROI)
- Internal rate of return (IRR)
- Net present value (NPV)
- Net operating income (NOI)
- Impact on asset value

In addition to the above outputs, the calculator also estimates the impact the proposed energy efficiency changes will have on a property's ENERGY STAR score.

The tool provides two ways to use its calculations. Users can save and print a summary of their results, or they can generate a letter highlighting the financial value for use as part of a capital investment proposal. Because energy efficiency projects generally improve net operating income, which is an important consideration when buying and selling commercial properties, the Financial Value Calculator provides strong financial arguments in favor of implementing these projects.

NEW Financing Programs Decision Guide for State and Local Governments

ENERGY STAR recently published the Financing Programs Decision Guide (http://www.epa.gov/statelocalclimate/state/activities/guide.html). This guide will help state and local governments design the appropriate finance programs for their jurisdictions, both public and private. Financing strategies for both energy efficiency and renewable energy are covered. The document is written for users with limited financial background.

The Decision Guide describes:

* Financing-program options
* Key components of successful programs
* Factors for states and communities to consider as they make decisions about getting started or updating their programs

This Decision Guide helps state and local governments offer financing support in the commercial and residential sectors. It also helps them finance improvements to their own buildings.

Chapter one of the Decision Guide outlines how the first steps in choosing a financing program involve decisions about the objectives, and helps the reader understand the requirements to ensure success.

Chapter two addresses the five key elements of a financing program, including: the target market, funding sources, security, credit enhancements, and loan origination and servicing.

Chapter three is a guide to the online Financing Program Decision Tool, which helps state and local governments choose a financing program that best fits their circumstances. The Tool helps users narrow down potential options by asking nine questions about preferred target markets and available resources. Once the user has honed in on applicable options, clicking on a selected option leads to a detailed description of that option, including case histories and links to other resources.

Chapter four, the last chapter, contains financing tools and resources that are most helpful to program designers. The chapter includes analytical tools, identifies funding databases, guidance, advisory services and toolkits, webinars, and successful marketing programs.

Cash Flow Opportunity Calculator, Version 2.0

The Cash Flow Opportunity (CFO) Calculator has proven to be a very effective tool, especially for public-sector projects. This set of spreadsheets helps create a sense of urgency about implementing energy efficiency projects by quantifying the costs of delaying the project. It was developed to help decision makers address three critical questions about energy efficiency investments:

1. How much of the new energy efficiency project can be paid for using the anticipated savings?

2. Should this project be financed now, or is it better to wait and use cash from a future budget?

3. Is money being lost, and how much, by waiting for a lower interest rate?

Using graphs and tables, the CFO Calculator is written so that managers who are not financial specialists can make informed decisions, yet it is sophisticated enough to satisfy financial decision makers. The updated tool works well for all types of projects in all market sectors (municipal, commercial, federal, residential including single family homes, etc.).

To determine how much of the new project can be paid for using your anticipated savings, the CFO Calculator takes a practical look at your energy efficiency situation and financing opportunities. The new CFO Calculator provides an easy interface that will accommodate energy efficiency retrofits in a variety of marketing situations: individual or groups of buildings, waste water facilities, manufacturing facilities, technology improvements, and USGBC's LEED projects. Whichever interface you choose, the CFO Calculator provides answers to some critical financial questions in just minutes.

The first step in this process is to estimate the amount of the savings that can be captured from the existing utility budget. The working assumption is that these savings will be used to cover the financing costs, and that the savings will recur. The savings amount is entered into a "reverse financial calculator," which then asks for an estimated borrowing interest rate, financing term, and the percentage of the savings you wish to use. It then calculates the amount of project improvements that could be purchased by redirecting these energy net savings to pay for the upgrades. When future energy savings are the main source of the project's repayment, the CFO Calculator becomes an effective sensitivity analysis tool that takes into account the impact of lower interest rates, longer financing terms, and utilization of savings when structuring the project's financing.

In one case study, the "See how much money you are leaving on the table" argument was made to the CFO of a large city in the Northeast on behalf of the local electric utility. The CFO responded that the city was fiscally conservative and believed that waiting until funds were available in a future operating budget (thereby avoiding borrowing and paying interest) was in the best interests of the city. The CFO Calculator was used to map the cash flow consequences of these two decision points (financing now or waiting until a future budget) to demonstrate to the city's CFO and city council that *financing now* was a better financial decision than waiting for cash. In fact, in most instances, the lost energy savings incurred by waiting for one year are greater than the net present value of all the interest payments of most financing, making "do it now" the better financial decision. This is counter intuitive and surprises most decision makers. Today, this city supports the expeditious implementation of energy efficiency projects.

Another common argument for delay is waiting for a lower interest rate offering rather than financing at a higher rate that is available immediately. This situation may occur when waiting for funds from a future bond issue or for a low-cost specialty fund to replenish itself, versus accepting an immediately available third-party financing offering. The CFO Calculator allows you to compare two different interest rate offerings, and it will compute how long you can wait for the lower interest rate before the lower rate begins to cost more. It does this by including the forfeited energy savings into the decision making process, which is truly another "cost of delay."

For more information if you are interested in understanding the underlying logic and math used in this tool, please read "A Look Inside the Cash Flow Opportunity (CFO) Calculator: Calculations and Methodology" which can be found on the energystar.gov Web site.[4]

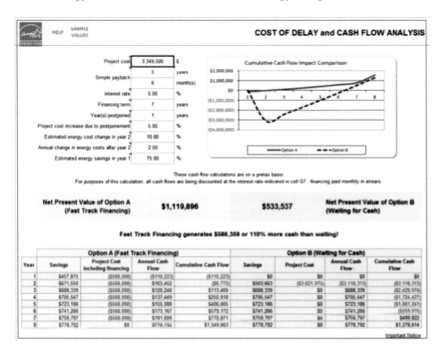

Figure 3-1. This is a sample screen capture of the "cash flow" tab from the CFO Calculator Excel Spread Sheet supporting the "do it now" argument.

CHOOSING THE RIGHT FINANCING

"Financing" should be thought of as a two-step process: financial instruments (or vehicles) and sources of funds. Once you decide which financing vehicle is best for your organization, the next step is to choose the best source of funds. Bear in mind that no one financing alternative is right for everyone. In the world of energy efficiency finance, one size definitely does not fit all! For the purposes of this discussion, our focus is limited to the public and private sectors, not consumer finance or federal financing options.

Before you can choose the right financial vehicle, however, two other issues must be considered tax-exempt status and interest rates. Public sector organizations and some non-profits qualify for tax-exempt financing, while private sector organizations generally do not. Private sector organizations are usually driven by tax considerations and financial strategies, but public sector organizations do not pay taxes and therefore may not be able to take advantage of tax benefits.

"What's the interest rate?" is frequently the first question asked when evaluating financing options. Organizations able to issue tax-exempt obligations will benefit more from lower interest rates than would be the case for regular "for profit" organizations. This is because, according to the Internal Revenue Code of 1986, the lender does not have to pay federal income tax on the interest earned from tax-exempt transactions. Due to competitive market forces, much of this saving is passed back to the borrower in the form of lower interest rates. Any U.S. state, district, or any subdivision thereof that (a) has the ability to tax its citizens; (b) has police powers; or (c) has the right of eminent domain can issue tax-exempt financing. This includes public schools, state universities and community colleges, libraries, public hospitals, town halls, municipal governments, county governments—in summary, almost any organization that receives its funding from tax revenues. While not-for-profit organizations created under Section 501(c)(3) of the Internal Revenue Code and private organizations do not directly qualify as issuers of tax-exempt obligations, they may be able to have a "conduit agency"—a city, state, health, or education authority—apply for the financing on their behalf.

For private sector organizations, interest rate alone is rarely the best indicator of the "best deal." To show the importance of proper deal structuring, consider the following question: "Which is the better finance offering, 0% or 6%?" Most people say "0%" until they find out that the 0% obligation must be repaid in 6 months, while the 6% obligation has a term of 5 years, making *cash flow* the deciding factor. The question is further complicated when the 6% obligation requires that the owner(s) personally guarantee the financing; however, the transaction can be done at 12% without a personal guarantee. Once all of the terms and conditions are known, it is easier to understand why many would prefer to pay a 12% financing interest rate rather than 0%. In addition to the term and personal guarantees, the list of structuring points is broad and may include whether the rate is fixed or variable, payment schedules, down payments, balloon payments, balance sheet impact, tax treatment, senior or subordinated debt, and whether additional collateral is required.

Your organization's legal structure, size, credit rating, time in business, sources of income, and profitability or cash flow are also important considerations when choosing the right financing vehicle and source of funds. The type of project, general market conditions, dollar size of the project, and the use of the equipment being financed are also important.

FINANCIAL INSTRUMENTS

There are two basic approaches to funding projects, "pay-as-you-go" and "pay-as-you-use." Pay-as-you-go means paying for the project out of current revenues at the time of expenditure; in other words, paying cash. If you don't have the cash, the project gets postponed until you do. Pay-as-you-use means borrowing to finance the expenditure, with debt service payments being made from revenues generated during the useful life of the project. Because energy efficiency projects generate operating savings over the life of the project, the pay-as-you-use approach makes good sense.

As previously mentioned, public-sector organizations can borrow at tax-exempt rates, which are substantially lower than the taxable rates that private sector organizations will pay when financing. While tax-exempt financial instruments can only be used by public sector (and qualifying non-profit and private sector) organizations, taxable instruments can be used by all. This section will help identify the financial instruments best suited for public- or private-sector application. Public-sector instruments include bonds and tax-exempt lease purchase agreements. Private-sector instruments include a variety of commercial leases. All sectors can benefit from *energy performance contracts* and *power purchase agreements*. Bear in mind that there are exceptions to every rule, and structuring financing to comply with tax or budget issues is often complicated, which is why working with a financial advisor often proves helpful.

Major capital projects are funded by some form of *debt*, which is categorized as either short term (for periods of less than twelve months) or long term (for periods greater than twelve months). Most borrowings by public-sector organizations require citizen approval, either directly through referendum or indirectly through actions of an appointed board or elected council. However, revenue bonds and tax-exempt lease-purchase agreements may not require local voter approval. (See details below.)

Most of us are familiar with traditional *loans*, which are debt obligations undertaken by a borrower. The title of the asset being financed is typically in the name of the borrower, and the lender files a lien on the asset being financed, prohibiting the borrower from selling the asset until the lien is lifted. Banks frequently require additional collateral, which may take the form of keeping compensating balances in an account or placing a blanket lien on all other assets of the organization. A *conditional sales agreement* or installment purchase agreement is a kind of loan that is secured by the asset being financed; the title of the asset transfers to the borrower after the final payment is made. All loans are considered "on balance sheet" transactions and are common to both the private and public sectors.

Frequently used *short-term debt instruments* include bank loans (term loans or lines of credit), anticipation notes (in anticipation of bond, tax, grant or revenues to be received), commercial paper (taxable or tax-exempt unsecured promissory notes that can be refinanced or rolled over for periods exceeding one year), and floating rate demand notes (notes that allow the purchaser to demand that the seller redeem the note when the interest rate adjusts).

Long-term debt is frequently in the form of *bonds*. Commercial bonds can be quite complex (asset backed, callable, convertible, debenture, fixed or floating rate, zero coupon, industrial development, tax credits, etc.) and usually require working with an investment banker. In the public sector, municipal bonds fall into two basic categories, general obligation (GO) bonds, and revenue bonds. *GO bonds* are backed by the issuer 's full faith and credit and can only be issued by units of government with taxing authority. Because the issuer promises to levy taxes to pay for these obligations if necessary, these bonds have the lowest risk of default and, therefore, the lowest cost.

Interest paid on GO bonds is typically exempt from federal income taxes and may be exempt from state income taxes.

Revenue bonds are also issued by local governments or public agencies. However, because they are repaid only from the specific revenues named in the bond, they are considered to be riskier than GO bonds. Revenue bonds may not require voter approval and often contain covenants intended to reduce the perceived risk. Typical covenants include rate formulas, the order of payments, establishing sinking funds, and limiting the ability to issue new debt. Small municipalities that have difficulty issuing debt often add credit enhancements to their bonds in the form of bond insurance or letters of credit.

Tax Credit Bonds. The American Recovery and Reinvestment Act of 2009 (ARRA) offers states, local governments, and conduit borrowers new financing options. Three specific options are Qualified School construction Bonds (QSCBs), Qualified Energy Conservation Bonds (QECBs), and New Clean Renewable Energy Bonds (NCREBs). These Tax Credit Bonds permit the bond holders to capture a federal tax credit, the amount of which is determined by the Secretary of the Treasury. This effectively reduces the net interest costs to the borrower to below market rates; often less than three percent (in today's market). Because the poor economy has limited many investors' ability to utilize tax credits (to offset earnings), tax credits may be converted into grants and used to reduce the investment needed to install the proposed equipment. The authorized amounts, accessibility, and issuing requirements can be complex, and are addressed by ARRA and the Internal Revenue Service.

In the case of most energy efficiency projects, the source of repayment is the actual energy savings (considered part of the operating budget) realized by the project. When the approval process to obtain the necessary debt is a barrier, public-sector organizations may be able to limit the repayment of the financing costs to their operating budget by using a tax-exempt lease purchase agreement. This solution may avoid the capital budget process altogether.

Tax-exempt Lease-purchase Agreements

Tax-exempt lease-purchase agreements are the popular public-sector financing alternatives that are paid from operating budget dollars rather than capital budget dollars. A tax-exempt lease purchase agreement is an effective alternative to traditional debt financing (bonds, loans, etc.), because it allows a public organization to pay for energy upgrades by using money that is already set aside in its annual utility budget. When properly structured, this type of financing makes it possible for public-sector agencies to draw on dollars to be saved in future utility bills to pay for new, energy efficient equipment and related services today.

A tax-exempt lease-purchase agreement, also known as a Municipal Lease, is closer in nature to an installment-purchase agreement than a rental agreement. Under most long-term rental agreements or commercial operating or true leases (such as those used in car leasing), the renter or lessee *returns* the asset (the car) at the end of the lease term without building any equity in the asset being leased. In contrast, a lease-purchase agreement presumes that the public sector organization will *own* the assets after the term expires. Further, the interest rates are appreciably lower than those on a taxable commercial lease-purchase agreement because the interest paid is exempt from federal income tax for public sector organizations.

In most states, a tax-exempt lease-purchase agreement usually does **not** constitute a long-term "debt" obligation because of non-appropriation language written into the agreement and, therefore, rarely requires public approval. This language effectively limits the payment obligation to the organization's current operating budget period (typically a 12-month period). The organization will, however, have to assure lenders that the energy efficiency projects being financed are considered of *essential use* (i.e., essential to the operation of the organization), which minimizes the nonappropriation risk to the lender. If, for some reason, future funds are not appropriated, the equipment is returned to the lender, and the repayment obligation is terminated at the end of the current operating period without placing any obligation on future budgets.

Public-sector organizations should consider using a tax-exempt lease-purchase agreement to pay for energy efficiency equipment and related services when the projected energy savings will be greater than the cost of the equipment (including financing), especially when a creditworthy energy service company (ESCO) guarantees the savings. If your financial decision makers are concerned about exceeding operating budgets, they can be assured this will not happen, because lease payments can be covered by the dollars to be saved on utility bills once the energy efficiency equipment is installed. Utility bill payments are already part of any organization's normal year-to-year operating budget. Although the financing terms for lease-purchase agreements may extend as long as 15 to 20 years, most lenders prefer to keep them to 12 years or less, and are limited by the useful life of the equipment.

There may be cases, however, when tax-exempt lease-purchase financing is not advisable for public sector organizations; for example, when (1) state statute or charter may prohibit such financing mechanisms; (2) the approval process may be too difficult or politically driven; or (3) other funds are readily available (e.g. bond funding that will soon be accessible or excess money that exists in the current capital or operating budgets).

How is Debt Defined in the Public Sector?

It is important for managers to be aware of the different interpretations of "debt" from three perspectives—legal, credit rating, and accounting. As mentioned above, most tax-exempt lease-purchase agreements are not considered "legal debt," which may prevent the need to obtain voter approval in your locality. However, credit rating agencies, such as Moody's Investor Services, Standard & Poor's, and Fitch Ratings Ltd., do include some or all of the lease-purchase obligations when they evaluate a public entity's credit rating and its ability to meet payment commitments ("debt service"). These two perspectives (legal and credit rating) may differ markedly from the way lease-purchase agreements are treated (i.e., which budget is charged) by your own accounting department and your organization's external auditors.

In general, lease-purchase payments on energy efficiency equipment are small when compared to the overall operating budget of a public organization. This usually means that the accounting treatment of such payments may be open to interpretation. Because savings occur only if the energy efficiency projects are installed, the projects' lease-purchase costs (or the financing costs for upgrades) can be paid out of the savings in the utility line item of the operating budget. Outside auditors may, however, take exception to this treatment if these payments are considered "material" from an accounting perspective. Determining when an expense is "material" is a matter of the auditor's professional judgment.[5] While there are no strictly defined accounting thresholds, as a practical guide, an item could be considered material when it equals or is greater than 5% of the total expense budget in the public sector (or 5% of the net income for the private sector). For example, the entire energy budget for a typical medium-to-large school district is around 2% of total operating expenses; therefore, so long as the payments stay under 2%, energy efficiency improvements will rarely be considered "material" using this practical guideline.

What are Energy Performance Contracts?

In most parts of the United States, an energy performance contract (EPC) is a common way to implement energy efficiency improvements. It frequently covers financing for the needed equipment, should your organization chose not to use internal funds. In fact, every state except Wyoming[6] has enacted some legislation or issued an executive directive to deal with energy efficiency improvements using EPCs. While EPCs are used both in the public and private sectors, 82% of the revenues of ESCOs come from public sector clients.[7] Properly structured EPCs can be treated as an operating, rather than a capital expense.

If you search for the phrase "energy performance contract," a variety of definitions appear:

- The U.S. Department of Housing and Urban Development (HUD) says that a performance contract is "an innovative financing technique that uses cost savings from reduced energy consumption to repay the cost of installing energy conservation measures..." The Energy Services Coalition—a national non-profit organization composed of energy experts working to increase energy efficiency and building upgrades in the public sector through energy savings performance contracting states that a performance contract is "an agreement with a private energy service company (ESCO)... [that] will identify and evaluate energy saving opportunities and then recommend a package of improvements to be paid for through savings. The ESCO will guarantee that savings meet or exceed annual payments to cover all project costs... If savings don't materialize, the ESCO pays the difference... *Notice that both definitions mention "payment." Let's dig a little deeper.*

- DOE claims that "Energy performance contracts are generally *financing* or *operating leases* provided by an Energy Service Company (ESCO) or equipment manufacturer. What distinguishes these contracts is that they provide a guarantee on energy savings from the installed retrofit measures, and they usually also offer a range of associated design, installation, and maintenance services."

- The NY State Energy Research and Development Authority (NYSERDA) states that "An EPC is a method of implementation and project financing, whereby the operational savings from energy efficiency improvements is amortized over an agreed-upon repayment period through a *tax-exempt lease purchase arrangement...*"

- Meanwhile, the Oregon Department of Energy says, "An energy savings performance contract is an agreement between an energy services company (ESCO) and a building owner. The owner uses the energy cost savings to reimburse the ESCO and to pay off the *loan* that financed the energy conservation projects."

Although the definition of a performance contract is not universally defined, it usually includes a variety of services, including: energy audits; designing, specifying, selling, and installing new equipment; providing performance guarantees; maintenance; training; measurement and verification protocols; financing; indoor air quality improvements, and more. One major benefit of using a performance contract is the ability to analyze the customer's needs and craft a custom agreement to address the organization's specific constraints due to budget, time, personnel, or lack of internal expertise. This includes choosing the financing vehicle that best suits the organization's financial and/or tax strategies.

Designed for larger projects, performance contracting allows for the use of energy savings from the operating budget (rather than the capital budget) to pay for necessary equipment and related services. Usually there is little or no up-front cost to the organization benefitting from the installed improvements, which then frees up savings from reduced utility bills that would otherwise be tied up in the operating budget. An energy performance contract is an agreement between the organization and an ESCO to provide a variety of energy saving services and products. Because these improvement projects usually cover multiple buildings and often include upgrades to entire lighting and HVAC systems, the startup cost when *not* using an EPC may be high, and the payback period may be lengthy. Under a well-crafted EPC, the ESCO will be paid based on the verifiable energy savings.

The ESCO will identify energy saving measures through an extensive energy audit and then install and maintain the equipment and other upgrades. This includes low- and no-cost measures which contribute to the project's overall savings. The ESCO works closely with the client throughout the approval process to determine which measures to install, timing of the installations, staffing requirements, etc. The energy savings cover the costs of using the ESCO *and* financing for the project.

TYPES OF ENERGY PERFORMANCE CONTRACTS

The most common type of performance contract is called a *"Guaranteed Savings Agreement,"* whereby the ESCO guarantees the savings of the installed energy-efficiency improvements (equipment and services). The ESCO assumes the performance risk of the energy efficient equipment so that if the promised savings are not met, the ESCO pays the difference between promised savings and actual savings. If the savings allow, a performance contract may include related services, such as the disposal of hazardous waste from the replacement of lighting systems or from the removal of asbestos when upgrading ventilation systems. The ESCO usually maintains the system during the life of the contract and can train staff to assist or continue its care after the expiration of the contract period. The ESCO can also play a major role in educating the customer organization about its energy use and ways to curb it.

A *shared savings agreement* is another type of energy services performance contract under which the ESCO installing the energy-efficient equipment receives a share of the savings during the term of the agreement. In a *fixed shared savings agreement*, the customer agrees to a payment based on stipulated savings, and once the project is completed, the payments usually cannot be changed. After the completion of the project, the savings are verified by an engineering analysis or other mutually agreed upon method. In a *true shared savings agreement*, the savings are verified on a regular basis, with the savings payments changing as the savings are realized.

In summary, performance contracts typically contain three identifiable components: a *project development agreement* indicating which measures will be implemented to save energy (and money); an energy services agreement indicating what needs to be done after the installation to maintain ongoing savings; and a *financing agreement*.[8] Organizations may choose to finance the projects independently of the ESCO, especially when they can access lower-cost financing on their own (as in the case of public sector organizations when accessing tax-exempt funding). So, what is the funding mechanism used in an EPC? Is it a financing or operating lease (two very different structures), a tax-exempt lease-purchase agreement, or a loan? From a financing perspective, these are all different vehicles with diverse accounting and tax consequences. The answer is *yes* to all.

Managed Energy Service Agreements (MESAs) are also being used to finance energy efficiency improvements. In a MESA, the energy user pays the energy efficiency supplier based on their historic energy use, and the supplier pays the utility bill directly. The energy efficiency supplier makes the energy efficiency improvements and uses the energy optimization savings to cover the cost of the new equipment and their services. Because the new equipment is owned and maintained by a special purpose entity established for this purpose, the building owner may be able to keep this financing off their balance sheet. In addition, MESAs may address the "split incentives" hurdle (tenants benefit from a reduced utility bill, while the building owner takes on the debt) which often keeps land-lords from implementing energy efficiency projects; the MESA sub-charges can be passed through to the tenants, just like their utility bills.

Regardless of the type of energy services agreement, it is important to remind the reader of two critical components that are needed to ensure that the energy performance and operational goals are met: (1) commissioning, and (2) measurement and verification. Commissioning is the process of making sure a new building functions as intended and communicating the intended performance to the building management team. This usually occurs when the building is turned over for occupancy. Ongoing and carefully monitored measurement and verification protocols are vital to ensure the continuing performance of the improvements, especially when the energy savings are the source of the financing repayment.

Power purchase agreements (PPAs), also known as design-build-own-operate agreements, are ones in which the customer purchases the measurable output of the project (e.g., kilowatt hours, steam, hot water) from the ESCO or a special purpose entity (SPE) established for the project, rather than from the local utility. Such purchases are at lower rates or on better terms than would have been received by staying with the utility. These agreements work well for on-site energy generation and/or central plant opportunities. PPAs are frequently used for renewable energy and cogeneration projects (also known as combined heat and power projects). Due to the complexities of the contracts, projects using PPAs are typically large. PPAs are frequently considered "off balance sheet" financing and are used in both the public and private sectors.

Commercial Leasing

Energy efficiency equipment that is considered by the Internal Revenue Service (IRS) as personal property (also known as "movable property" or "chattels") may be leased. The traditional equipment lease is a contract between two parties in which one party is given the right to use another party's equipment for a periodic payment over a specified term. Basically, this is a long-term rental agreement with clearly stated purchase options that may be exercised at the end of the lease term. Commercial leasing is an effective financing vehicle and is often referred to as "creative financing." Leases can be written so the payments accommodate a customer 's cash flow needs (short-, long-, or "odd-" term; increasing or decreasing payments over time; balloon payments;

skip payments, etc.). Leases are frequently used as part of an organization's overall tax and financing strategy and, as such, are mostly used in the private sector.

From a financial reporting perspective, however, commercial leases fall into only two categories, an *operating lease* or a *capital lease*. Each has substantially different financial consequences and accounting treatment. The monthly payments of an *operating lease* are usually lower than loan payments because the asset is owned by the lessor ("lender"), and the lessee's ("borrower's") payments do not build equity in the asset. The equipment is used by the lessee during the term, and the assumption is that the lessee will want to return the equipment at the end of the lease period. This means that the lease calculations must include assumptions that the residual value of the leased asset can be recovered at the end of the lease term. In other words, equipment that has little or no value at the end of the lease term will probably not qualify under an operating lease. For example, lighting systems would not qualify, while a well maintained generator in a cogeneration project might. Operating leases are considered "off balance sheet" financing, and payments are treated as an operating expense.

A common *capital lease* is a "finance lease," which is similar to a conditional sales agreement because the asset must be reflected on the lessee's (borrower's) balance sheet. A finance lease is easily recognized because the customer can buy the equipment at the end of the lease term at a stated price that is less than its fair market value ("a bargain purchase option"). For example, a lease with a one dollar purchase option is clearly a capital lease. Other conditions that define a capital lease deal with the term of the lease, transfer of ownership, and lessor's equity in the asset.[9]

Not everyone realizes that the tax treatment of a lease may be different from the financial reporting treatment of a lease. A *tax lease* or *guideline lease* is one in which the lessor keeps the tax incentives provided by the tax laws for investment and ownership of equipment (typically depreciation and tax credits). Generally, the lease rate on tax or guideline leases is reduced to reflect the lessor's recognition of this tax incentive. A *true lease*, similar to a long-term rental agreement, gives the lessee the option to buy the equipment at its true fair market value at the end of the lease term and may allow the lessee to deduct the monthly lease payments as an operating expense for income tax purposes. After all, you can't depreciate an asset that you do not own. However, a true lease will be picked up on the balance sheet.

It is important to note that these leasing definitions are in effect as of the revision of this document (February, 2012). However, in an effort to create financial transparency for investors, the Financial Accounting Standards Board (FASB) and the International Accounting Standards Board (IASB) are currently evaluating alternative accounting treatments for *operating leases*. At this time, it appears that the distinction between *operating* and *capital leases* will disappear. If approved, companies will have to reflect all assets they have the right to use by creating a new balance sheet category "right of use" asset (determined by the Net Present Value of the lease payments), which would be offset by a corresponding liability. Currently, operating lease information is found in the Notes to Financial Statement and is not reflected on the Balance Sheet. The financial community expects this change to have a negative effect on companies with existing loan covenants, requiring lenders to renegotiate the covenants or call in the loan(s). The real estate industry is concerned that renters will have to reflect their leases on their balance sheet, which could hurt the industry. The final guidance document is expected to be published at the end of 2012, with implementation delayed until 2015.

Public-sector organizations frequently lease equipment. However, because most public-sector organizations are tax-exempt, tax strategies are not usually a consideration when deciding which type of lease to enter into.

Commercial PACE.

PACE stands for Property Accessed Clean Energy. Essentially, it provides the ability to pay for energy efficiency improvements and renewable energy projects by creating a special assessment added to the property owners' property taxes for up to 20 years. It is modeled after a familiar structure used to finance water and sewer system hookups, roads, parks, etc. In essence, PACE ties the repayment of the loan to the property rather than the owner, allowing for longer term financing and low interest rates (due to the secured interest in the real property). Unfortunately, the Federal Housing Finance Authority raised issues about providing PACE financing for residential properties, effectively stopping the market for residential PACE However, commercial properties are not subject to the same regulations, and a number of cities are moving forward with this innovative alternative financing plan. For more and up-to-date information about PACE, please check www.pacenow.org.

FUNDING SOURCES

Once you have determined that internal sources of funds are not available or are insufficient for your energy efficiency project, your options become (a) using third party lenders, (b) postponing the project, or (c) installing part of the upgrades by breaking the project into smaller pieces. Earlier in this chapter, we explained why postponing or delaying the project can, in fact, be the most expensive alternative. Consequently, financing often becomes the best decision. Once you decide to finance the project and identify a preferred financing vehicle, the next step is to evaluate potential funding sources.

Traditional funders include banks, commercial credit companies, leasing companies, insurance companies, brokerage houses, and vendors. If you are dealing with a large financial institution, it is important to contact the right department within that organization to obtain the best pricing for your project. For example, when speaking to their bank, most people start with their "regular banker," who may limit their discussions to loans. Keep in mind, however, that most larger banks have a public finance department where you will find tax-exempt lease purchasing, and sometimes even a bank-owned leasing company where you may structure a special equipment lease. Large commercial credit companies may divide the market by the size of the transaction—small ticket, middle market, and large ticket. To further complicate matters, companies often define their market divisions differently (i.e., under $25k for micro-ticket, $25k-$500k for small ticket, $500k-$15 million for middle market, and over $15 million for large ticket[10]). If your project is $500,000, the small ticket and middle-market groups may quote you two different prices, even though they work for the same lender. Add the dimension of public-sector versus private-sector finance, and you may get different pricing again. This is especially true in organizations where the sales staff's compensation plan is commission based, and they do not have a company lead-sharing policy in place.

In almost every state where the electric industry has been restructured (deregulated), legislation has been passed to create a system benefits charge (also known as a public benefits charge) that adds a defined surcharge (fee) to the electricity bills. These fees are used to support energy-related projects that provide public benefits such as renewable energy, energy efficiency, low-income customer programs, energy R&D, or other related activities that may not be available in a competitive market. These fees, usually a per-kilowatt (kWh) hour cost or a fixed charge, are charged to all customers and cannot be bypassed. The amount of the charge varies by state and accumulates in a fund, which is usually administered by the state's energy office or local utility.

If you are located in a state with public benefits charges (see map below), you may find that your energy efficiency project qualifies for a low-cost or below-market financing program funded by these charges. A listing of the total funding amounts by category and administrative contact information is available at the American Council for an Energy-Efficient Economy's website http://www.aceee.org/sector/state-policy/utility-policies. In the regulated states, the public utility commissions continue to provide incentives (rebates, loans, grants, etc.) to promote energy efficiency and renewable energy projects.

Examples of some of these programs include Texas' LoanSTAR Revolving Loan Program (available to schools, local governments, state agencies, and hospitals), which will loan up to $5 million at 2.5% interest (today's rate). In Oregon, the Ashland Electric Utility offers 0% loans to their commercial customers, nonprofit and local governments for up to $10,000 to finance energy efficiency improvements to their facilities. The loans can be used for lighting retrofits and other energy-efficiency measures. Connecticut offers a variety of programs for residential-, commercial-, and public-sector organizations. For example, their Small Business Energy Advantage Program will finance up to $100,000 for private-sector, and $400,000 for public-sector customers at 0% interest for terms up to 36 months. The programs are funded by **The Clean Energy Finance and Investment Authority (CEFIA)**—formerly The Connecticut Energy Efficiency Fund—and administered by **CL&P** and **UI**.

Virtually every state offers some form of incentive, and we recommend starting your funding search by reviewing these special programs. A good place to find a list of energy efficiency incentives is the Database of State Incentives for Renewables and Efficiency (www.dsireusa.org).

VISION OF THE FUTURE

In the United States, energy efficiency is big business. A recent Lawrence Berkeley National Laboratory-National Association of Energy Service Companies (LBNL-NAESCO) survey states, "The ESCO industry continues to grow with estimated revenues of $4.1 billion in 2008 despite a general downturn in the broader economy" with "ESCO industry in ag-

2011 ACEEE State Energy Efficiency Scorecard Map

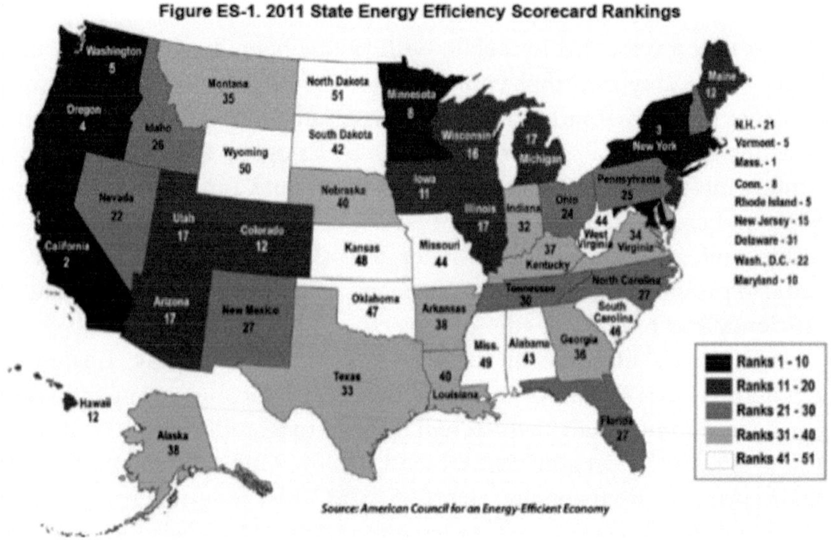

Figure ES-1. 2011 State Energy Efficiency Scorecard Rankings

Source: American Council for an Energy-Efficient Economy

Figure 3-2.

NOTE: Light gray states have PBFs that support energy efficiency and renewable energy. Dark gray states have PBFs that support only energy efficiency (ACEEE 2004; UCS 2004).

gregate will have annual revenues of $7.1–7.3 billion in 2011."[11]

The survey went on to confirm that "MUSH" markets—municipal and state governments, universities and colleges, K-12 schools, and hospitals—have historically represented the largest share of ESCO industry activity, which was 69% of industry revenues in 2008. The federal market represented 15%, and public housing represented 3% of industry revenues, while the commercial and industrial sectors represented 7% (a drop from 15% in 2006). Energy is the single largest operating expense in an office building, representing about 30% of a typical building's costs. This means that the commercial sector represents a growth opportunity for ESCOs.

Historically, however, the commercial real estate sector has faced numerous barriers to increasing their energy efficiency efforts, including:

• Split incentives (when the tenant, not the building owner, pays the utility bills).

- Properties are held under complex legal structures that make it difficult and time consuming to obtain financing (limited liability companies, limited partnerships, etc.).

- Non-recourse financing may already be in place, but new lenders want recourse on traditional energy efficiency project financings, especially when the equity in the property is highly leveraged.

- Difficulty in obtaining a security interest in the new energy assets being installed unless the entire building is refinanced.

- Tenants not wanting to take on debt for long-term leasehold improvements.

- Lenders not wanting to provide secondary financing to building owners for terms longer than the remaining lease term for key tenants.

- Properties are often managed by professional, third-party management companies that generally cannot enter into debt obligations on behalf of owners without special authority.

- Building owners and/or tenants may be unable or unwilling to borrow more money, as they may be concerned about reaching their debt capacity or violating covenants in existing loan agreements.

While many responsible building owners have "stepped up to the plate" and committed to financing new energy efficiency projects, this market sector still represents considerable marketing opportunities and challenges. An innovative financing alternative may help in "getting to yes."

To address this marketing challenge, the adage "the past is the key to the future" comes to mind. Back in 1991, Pacific/Utah Power, the electric generation and distribution divisions of PacifiCorp, offered their retail customers an innovative type of efficiency program in which customers repaid the costs of their efficiency installations through monthly energy service charges on their electric bills. As part of the utility's demand-side management program, the utility had recourse to shutting off power if the customer defaulted. In essence, the electric meter became the "credit." The utility did more than 1,000 transactions, and eventually sold their loan portfolio to a major U.S. bank.

The many benefits of this program included not requiring the tenant or building owner to enter into a debt obligation (thereby overcoming aversions to borrowing), tying the repayment to the use of the equipment, allowing the building owner to acquire new energy-saving equipment at no direct cost, and having the tenants reduce their monthly utility bills without incurring debt.

The challenges to making a program such as this work include persuading the appropriate state public utility commissions to authorize a new tariff (energy service charge), finding a lender willing to underwrite the program, and insuring the installed equipment works as promised. Perhaps a "bill-to-the-meter" financing program could be considered to increase participation by this market sector in installing energy efficiency projects.

CONCLUSION

This chapter demonstrated how public and private sector organizations can redirect energy inefficiencies and waste from their current and future operating budgets in order to pay for the needed energy efficiency improvements today. Practical suggestions that support the urgency of implementing these projects have been outlined and show how to quantify the costs of delay. One section reviewed useful, field-tested ENERGY STAR tools and resources that are in the public domain and yours to use. Finally, potential sources of low- or no-cost funding for your projects were discussed, along with a variety of alternative financing vehicles.

Clearly, a decision not to install more energy-efficient equipment and implement related energy-saving measures is a decision to continue paying higher utility bills to the utility company. Using the captured energy savings to pay for the financing of improvements is recommended, essentially making them "self-liquidating" obligations. Since these energy efficiency projects pay for themselves over time, the bottom line is that choosing to delay energy improvements may be the wrong decision and cost more in the long term.

References
1. According to Barron's Dictionary of Accounting Terms, capital expenditures are "outlays charged to a long-term asset account. A capital expenditure either adds a fixed asset unit or increases the value of an existing fixed asset." Operating expenditures are costs "associated with the … administrative activities of the [organization]."

2. Note, however, off-balance-sheet transactions will be referenced in the footnotes to the organization's financial statements.
3. Source energy includes the energy consumed at the building itself—or the site energy—plus the energy used to generate, transmit, and distribute the site energy.
4. http://www.energystar.gov/ia/business/downloads/A_Look_Inside_the_Cash_Flow_Opportunity_Calculator_FINAL.pdf?1796-beb2
5. According to James Donegan, Ph.D. (accounting), Western Connecticut State University, an amount is "considered material when it would affect the judgment of a reasonably informed reader when analyzing financial statements."
6. †http://www.ornl.gov/info/esco/legislation/, however the Energy Services Coalition reports that Wyoming has entered into over $13 million in EPCs
7. A Survey of the U.S. ESCO Industry: Market Growth and Development from 2000 to 2006." Lawrence Berkeley National Laboratory, LBNL-62679. May 2007.
8. It is important to note that savings are measured in kWh and therms, and then translated into dollars at the current market price for electricity and natural gas.
9. See Financial Accounting Standards Board Statement of Financial Accounting Standards No. 13 for more information. Current FASB 13 distinctions between operating and capital leases are based on commercial law.
10. There is no industry standard, and individual lenders may set these break points differently than this example.
11. http://eetd.lbl.gov/ea/ems/reports/lbnl-3479e.pdf

Chapter 4

The Evolution of Performance Contracts As a Financial Solution

Shirley J. Hansen, Ph.D.

Since soaring energy prices and the growing need for energy efficiency hit the center of our radar screen in the 1970s, there has been growing recognition that using energy more efficiently is good for the economy and the environment. It's just good business. Yet survey after survey reveals that many organizations put off energy-efficiency work for one major reason: money.

Organizations either lacked the money, or those that have the funds are inclined to spend their money elsewhere. The reasons are legitimate: "We must use the money to invest in new production equipment. We must buy new math books. The payback is not short enough," etc., etc. For years many of us thought it was due almost entirely to the discomfort top management felt when the subject of "energy" was introduced. It certainly played a part, but in retrospect we now realize that other concerns were at play.

The horrible truth is that much of top management is not interested in ENERGY! They don't want to hear about gigajoules or British thermal units. CEOs and CFOs do not buy energy; they buy what it can do. How can they worry about the efficiency of something that is virtually non-existent in their lives? The only time it seems to reach their consciousness is when there is a shortage, a sudden power outage, or a spike in prices.

Many years ago, a dear friend who was in top management in a major corporation, gave me some sage advice, that is fundamental to our problem. He said, "Shirley, you folks must learn to fish from the fish's point of view."

To fish from the fish's point of view, we must first learn about the fish. It is key to recognize that the top management fish is interested in delivering promised results, be it education, patient care, or selling widgets. Second, we need to be aware that management is "facility blind." Management can walk the corridors but seldom see the facility until something goes wrong. Third, such details as "energy" are just noise—a small irritating noise for someone else to deal with.

If we are to get the "fish's" attention, we must talk their language and make the case in their terms. For them to bite, the bait on the hook must make energy efficiency (EE) a solid business opportunity.

It is essential that we provide our "fish": (1) an effective cost/benefit analysis procedure, which compares the net benefits of energy efficiency to increased production; (2) a new perspective of energy savings as a percentage of the bottom line; and (3) energy efficiency and conservation as a very cost-effective delivery system for meeting environmental responsibilities and mandates—a way to make money while reducing emissions.

Probably our biggest challenge is to remind top management, as forcefully as possible, that EE can be a self-funding endeavor. CEOs and CFOs have a tendency to compare energy investments to other business investments and fail to appreciate that no new money is required to do energy efficiency work. The money needed for energy investments is already in the budget—and it's being spent on wasted energy. The financing source for the EE investment is right there in avoided utility costs—money that will go up the smokestack creating more pollution every day that the energy efficiency measures are not taken.

All this, however, becomes much more palatable if the initial expenditures are from someone else's pocketbook—even better if there is someone out there that will guarantee that the funds to do the energy work will come out of avoided utility costs (the future energy savings from the project).

With such a backdrop, it is not surprising that performance contracting emerged as an attractive financing mechanism for energy work. Imagine the jubilation when someone figured out that future energy costs could finance the rebuilding or replacement of decrepit heating and cooling systems. And it just got better, because the money being set aside to fix those behemoths could now be used for something else.

The concept of performance contract is the same idea that stands behind commercial paper: the fact that a future obligation to pay already has value today. The future obligation in performance contracting is not derived from serving a debt but rather from the known and unavoidable cost of heating, cooling and illuminating buildings or fueling industrial processes.

Administrators around the world are burdened with obsolete, money-hungry schools, hospitals, etc. built in the low-energy-cost era of the 1950s and 1960s. Unfortunately, they cannot accept re-equipping these structures with the immensely improved energy-use technologies that have emerged in recent years. They feel compelled to continue to throw good money after bad to keep buildings in service.

By a kind of financial judo, performance contracting turns this grim prospect into an asset. Future utility bills can be discounted, while the commercial debt and the resulting cash flow can be used today for retrofits.

PERFORMANCE CONTRACTING IN RETROSPECT

In the late 1970s, Scallop Thermal, a division of Royal Dutch Shell, introduced the concept of using third-party financing to improve energy efficiency and cut operating costs in North America. Scallop offered to meet a Philadelphia hospital's existing energy services for 90 percent of its current utility bill. By upgrading the mechanical system and implementing energy efficient practices, Scallop was able to bring consumption well below the 90 percent level. Scallop both paid for its services and made a profit from this difference.

These early energy financing agreements generally were based upon each party receiving a percentage of the energy cost savings. The energy service company (ESCO) received a share to cover its costs and make a profit. The owner also received a share (as well as capital improvements) as an incentive to participate. Since each party received a share of the energy cost savings, this procedure became known as "shared savings."

During the life of the contract, the ESCO expected its percentage of the cost savings to cover all costs it had incurred, plus deliver a profit. This concept worked quite well as long as energy prices stayed the same or increased.

In the mid-1980s, however, energy prices began to drop. With lower prices, it took longer than predicted for the firm to recover its costs. Some firms could not meet their payments to their suppliers or financial backers. Companies closed their doors and, in the process, defaulted on their commitments to their shared savings customers. Some suppliers tried to recover costs from the building owners. Lawsuits were filed and "shared savings" nearly died a painful death.

From this tenuous thread, the "shared savings" industry in North America survived, but its character changed dramatically. Those supplying the financial backing and/or equipment recognized the risks of basing contract on future energy prices. Higher risks meant higher interest rates, if the money could be found. Insurance, which had been available to ESCOs for "shared savings," became a scarce commodity. By the late 1980s, shared cost savings agreements had shrunk to approximately 5 percent of the market.

From the end user's point of view, through the "shared savings" boom, it soon became apparent that the energy payments to the ESCO were unpredictable. All too often, the customers found themselves paying far more than expected for the opportunity. A customer, who had accepted a shared savings deal with $3.5 million in equipment, who was asked to pay 70 percent of a predicted $1 million annual savings for five years, assumed the total payment would be about $3.5 million for the acquired equipment and services. If, however, the savings were greater than expected or the price of energy went up, the costs could easily become $5-7 million for the same equipment. Payment procedures became confrontational. In the final analysis, too often the only real benefit of the shared savings transactions was the "off balance" sheet feature. This was attractive to customers who did not wish to incur more debt.

Out of the "shared savings" confusion of the 1970s-1980s evolved a new financing mechanism with the guarantee centered on the reduction in the amount of energy consumed and the value of the energy in dollar savings calculated at current billing rates. Typically, the projected dollar savings were guaranteed to cover any of the associated debt service obligations of the owner. To avoid the risks associated with falling energy prices, ESCOs began setting an energy floor price below which money guarantees would not apply.

A new term, "performance contracting," emerged that embraced all guaranteed energy efficiency financing schemes, including shared savings and guaranteed savings.

The initial attraction to performance contracting was the financing help. Through the years, the ESCO expertise has become equally attractive. For example, a 1988 survey found that the majority of US school administrators still thought work on the building envelope, (e.g., added insulation, double glazing of windows, etc.) was the most cost-effective energy savings measure to take. However, a U.S. Department of Energy analysis at the time showed that building envelope measures averaged the longest payback (over eight years) of those measures studied. In contrast, control measures paid back in roughly two years. In attempts to cut energy costs, school administrators were not investing money the most effective place. If fact, they were waiting eight years to get the return on investment from insulation that they could have achieved in two years from certain control measures.

Over the years, ESCOs have learned what works—and often at great expense what doesn't. We should profit from their experience. The earlier temptation to invest in elaborate control systems, for example, no longer exists. Today, experienced ESCOs know exactly how sophisticated a system should be in a given facility. That expertise can keep owners from investing valuable money in the wrong measures.

Performance contracting has matured into a viable and reliable way of doing business in North America and many countries around the world. In the United States and Canada, federal law allows—even encourages—federal agencies to use performance contracting to cut operating costs. Over 4,000 of the US school systems have had some type of performance contract, and these school systems have as many as 800 buildings each. Major manufacturers of energy-related equipment have ESCO services and divisions, and many regional ESCOs have grown from engineering firms, distributorships and mechanical contractors. In some parts of the world, there is a rush to get into the ESCO business. In fact, the concept has become so popular there is great fear in the industry that "wannabe ESCOs," often dubbed WISHCOs, are offering services before they fully understand the complexity of the process and are creating credibility difficulties for the industry.

In the mid-1990s, the International Energy Agency went on record at the Conference on Energy Efficiency in Latin America to support and encourage the ESCO industry. More recently the European Union has been actively promoting the concept through papers and conferences. The ESCO Europe Conference, sponsored in part by the EU, has become an annual event. More recently, the ESCO concept has become accepted in Asia. In late 2007, the Japanese and Chinese ESCO associates put together the second Asia ESCO Conference in Beijing. In fact, representatives of the various country ESCO associations met at the Beijing conference to see if they might create an international networking organization. Some version of performance contracting is now being practiced in all industrialized nations and most developing countries. The US Agency for International Development, the World Bank, Asian Development Bank, and the European Bank for Reconstruction and Development are all very active in fostering ESCO industries in developing countries.

Information relative to ESCO activity in various countries is available in *ESCOs Around the World: Lessons Learned in 49 Countries* (2009). An update on the industry titled *World ESCO Outlook* is available now. The publisher of both books is The Fairmont Press.

TYPICAL ESCO SERVICES

Through the years, ESCO services have become more varied. It has become a customer-driven industry, and the customer typically has a selection of ESCO services from which to choose. Services offered by an ESCO usually include:

- An investment grade energy audit to identify energy and operational savings opportunities, assess risks, determine risk management/ mitigating strategies, and calculate cost-effectiveness of proposed measures over time;

- Financing from its own resources or through arrangements with banks or other financing institutions;

- The purchase, installation and maintenance of the installed energy-efficient equipment (and possibly maintenance on all energy-consuming equipment);

- New equipment training of operations and maintenance (O&M) personnel;

- Training of O&M personnel in energy-efficient practices;

- Monitoring of the operations and energy savings so reduced energy consumption and operation costs persist;

- Measurement and savings verification; and

•   A guarantee of the energy savings to be achieved.

The popularity of performance contracting rests on the many benefits it delivers. The highlights of the benefits are addressed below.

CUSTOMER ADVANTAGES

The recipient of ESCO services can achieve many benefits, including:

•   an immediate upgrade of facilities and reduced operating costs— without any initial capital investment;

•   access to the ESCO's energy efficiency expertise;

•   positive cash flow (most projects generate savings that exceed the guarantee);

•   the opportunity to use the money that would have been used for required upgrades or replacements to meet other needs;

•   improved and more energy-efficient operations and maintenance;

•   several normal business risks are assumed by the ESCO, including the guaranteed performance of the new equipment for the life of the contract (not just through the manufacturer's warranty period);

•   a more comfortable, productive environment;

•   services paid for with the money that the customer would otherwise have paid to the utilities for wasted energy.

Performance contracting is probably the best and quickest way to be sure an organization is operating as efficiently as possible. It offers the customer a risk shedding opportunity; however, risks do still exist. Risk shedding always carries a price tag. It should be noted, however, that, as the customer moves to the more comprehensive options, he, or she, also receives more services, has less administrative burden, and will probably achieve greater savings persistence. The costs associated with the "right-hand side of the menu" can also be mitigated. The greatest benefit of performance contracting, however, could be the relative speed with which projects can be implemented, thus avoiding valuable dollars going up the smokestack while the customer tries to get more of the energy efficiency work out of his or her organization's limited resources.

In performance contracting, the risks shed by the owner are largely assumed by the ESCO. Effective performance contracting then becomes a matter of risk assessment and management by the ESCO. The more accurately the ESCO assesses the risks and the more effectively it manages/mitigates them, the greater the benefits to all parties. Since performance contracting risks are primarily managed through the project's financial structure, effective risk management presents a major point of differentiation among ESCOs.

ESCO RISK ASSESSMENT AND MANAGEMENT

Effective, experienced ESCOs have found a series of critical risk assessment components and management techniques, which include:

Pre-qualify the customer. Prequalification criteria should be established; the customer's organizational, technical and financial data should be carefully collected, validated, and evaluated.

Conduct an investment-grade audit. The typical energy audit assumes that conditions observed in the audit will stay the same for the life of the equipment and/or the project. An investment-grade audit (IGA) assesses administrative risks, operation and maintenance risks and the impact these risks will have on the project's savings over time. An IGA also considers the time value of money. Projected payback calculations, for example, discount the value of the dollars saved each year for the life of the project. A typical four-year payback, for example, can easily become 5+ years in the real value of dollars saved, thus changing the dynamics of the cost effectiveness calculations. Further, the investment-grade audit incorporates the cost of risk mitigation in payback calculations. It also recognizes that facilities and processes are critical portions of an organization's investment portfolio. The investment grade audit, therefore, offers the owner a guide in ways to enhance the work environment and the value of the portfolio. A list of measures, which save energy but do not address the workplace conditions, is no longer sufficient.

Establish sound base year data. Base year data are more than the average energy consumption over the last two or three years. They also consider existing conditions and what was happening in the facility or industrial process during that period to cause that consumption. Hours of occupancy, level of occupancy, run times, etc., all become critical issues that must be verified prior to project implementation and signed off on by both parties. How these variables are adjusted to establish an annually adjusted baseline for reconciliation purposes is critical.

Secure a solid contract fair to all parties. Sometimes ESCOs get a little too zealous in managing their risks. For example, the following paragraph, nearly a deal stopper, was put into a draft contract in caps:

IN NO EVENT SHALL [THE ESCO] BE LIABLE FOR ANY SPECIAL, INCIDEN-TAL, INDIRECT, SPECULATIVE, REMOTE OR CONSEQUENTIAL DAMAGES ARISING FROM, RELATING TO, OR CONNECTED WITH THE WORK, THE SUPPORT SERVICES, EQUIPMENT, MATERIALS, OR ANY GOODS OR SER-VICES PROVIDED HEREUNDER.

After months of discussion, the ESCO finally accepted that no customer with astute judgment would agree to such terms. The paragraph was finally dropped and the project moved ahead.

Implement Quality Measures

Customers too often insist on quantity instead of quality. One federal procurement officer reportedly told an ESCO the military base would rather have "15 Tempos than 10 Volvos." The agency may have reasons for needing 15 sets of wheels, but when the focus is energy, the ESCO must look at parts, repairs, maintenance and savings persistence.

Operations and Maintenance (O&M)

Too often overlooked, energy efficient O&M practices are absolutely vital to the success of a project. An evaluation of the schools and hospitals federal energy grants program (Institutional Conservation Program) after eight grant cycles revealed that in effective energy management programs up to 80 percent of the savings could be attributed to energy efficient O&M practices. If ESCOs are to guarantee the savings from a measure, they must look beyond the equipment to its "care and feeding" over the life of the contract. If the ESCO does not perform the maintenance, it must; (a) train the owner's staff; (b) police the O&M practices; and/or

(c) discount the predicted level of savings. In all cases the associated risk level must be evaluated, and the associated risk mitigation/management costs included in project cost calculations.

Measurement and Savings Verification (M&V)

Once the recommended measures are known and approved, the ESCO and owner should cooperatively establish the desired level of accuracy and the associated costs for M&V, that they are willing to have the project carry. The question is: How much accuracy are the parties willing to pay for? When M&V costs are included in the project, they cut into the savings and thereby diminish the amount of equipment and services that can be funded. But anytime money changes hands based upon savings, those savings need to be quantified in an acceptable manner. The International Performance Measurement and Verification Protocol (IPMVP) is the most broadly accepted means of measuring and verifying savings. The IPMVP guidelines are available through the organization's website: evo-world.org.

Project Management

The "business end" of the project, securing the savings, doesn't begin until the equipment is installed. Weak ESCOs tend to think of the project as "complete" once it is in the ground. Customers often think the ESCO is in business to sell equipment. ESCOs typically lose money during equipment installation but make up for it through the service and savings over the life of the contract. A good project manager is an incredibly valuable part of the ESCOs project delivery during the project implementation and in all the contract years.

When ESCOs have completed a thorough risk assessment and determined the procedures needed to manage those risks, they then discount the savings to adjust for risk exposure. In other words, they never guarantee 100 percent of the expected savings.

PERFORMANCE CONTRACTING:
A FINANCIAL TRANSACTION

Risk is managed through the financial structure. The ESCOs investment grade audit will provide the information necessary to establish how much can be guaranteed and how much must be held back as a risk cushion. The financial structure has essentially three components:

- the guarantee, which covers design, acquisition, installation and the cost of money;

- the ESCO fee for services performed;

- the positive cash flow (the savings in excess of the guarantee and ESCO fee).

The size of the ESCO fee will vary with the services provided and the risks perceived and assumed by the ESCOs. Some ESCOs put their fee in the guarantee package, others take it off the top of any savings that exceed the guaranteed amount, and still others do some of each. Owners should realize that an ESCO that offers the customer "all of the positive cash flow" has all its fee imbedded in the guarantee package. In such cases, the customer will pay finance charges and interest on the ESCO fees for the life of the contract.

Performance contracting has had its problems. Many of them in recent years, however, can be traced to the assumption by fledgling ESCOs and customers alike, that performance contracting is a technical procedure. It is not. Technical issues are important; they are the foundation upon which the project is based. On the other hand, lawyers would have you believe it's all managed through the contract. It is not. A good contract is simply the basis of a good project.

Make no mistake: Performance contracting is primarily a financial transaction. The ultimate performance contract decisions cannot be made in the boiler room; they must be made in the business office. Engineers and lawyers must also be in that business office when the decisions are made, but a strong financial voice is crucial.

Performance contracting is a simple idea but a complex process. If we are to get the maximum benefit out of performance contracting, it helps if financial officers have some understanding of engineering and legal issues; however, it is absolutely critical that lawyers and engineers understand and appreciate the fundamental role that finance plays in performance contracting.

Chapter 5

Power Purchase Agreements for Solar Electric Systems
Perhaps the Most Prolific Form of Performance Contract

Ryan Park and Jim Coombs

INTRODUCTION

Over the past five years, the power purchase agreement ("PPA") has emerged as an effective alternative approach to financing solar electric projects for businesses, nonprofits, and municipal government agencies in the United States. Federal government agencies have recently initiated several PPA projects, and we believe federal solar projects will increasingly utilize PPA solutions. In this chapter, we will explain what a PPA is, and more importantly, we will provide context for our readers to understand the fundamental factors that may enable them to utilize a PPA to meet their financial and operational objectives.

As a starting point, let us emphasize that we approach solar projects as *both* a **financial product** *and* an **energy solution**. In our experience, businesses and governments are intrigued by solar because if offers an alternative to generate energy in an environmentally friendly way, but in all cases they will bring a firm set of financial requirements in order to move forward with a project. The benefits of a solar project may be important to management and shareholders of a business, or to city councils, school boards, and local leaders, who value sustainability. Moreover, employees, customers, citizens, and regulators may recognize the importance of solar initiatives. However, business and government leaders are always motivated by financial best practices, and have a duty to all of their stakeholders—whether taxpayers or shareholders—to utilize their funds responsibly, and without subjecting them to unreasonable risks.

Fortunately, solar projects in general—and PPAs in particular—deliver quantifiable financial benefits and minimal risk when properly planned and implemented.

The solar PPA has been developed primarily in the US, as opposed to Europe and Canada, due to the unique "New Metering" ("NEM") approach that is broadly favored in America by utilities and state governments. Net metering is straightforward to understand, and is critical to understanding how PPAs work and how they deliver financial savings. Europe and Canada have predominantly favored the Feed-In Tariff ("FIT") as a means of incentivizing solar development, which presents a similar but distinctly different financial solution from a net metering regime.

The broader point in discussing differences between the United States and Europe, and between individual states and localities within the US, is that the financial incentives provided by federal, state, and in some cases local governments, as well as by utilities, are an indispensable component of any solar project, including a power purchase agreement. These incentives vary dramatically between states and countries. The reader should keep in mind that currently the ability to implement a PPA that will yield financial benefits for the host customer is dependent on incentives applicable within the utility serving each individual project site. For facilities owners with only one site, this can make or break the potential for a viable PPA. For those with multiple sites spread through various utilities and states, the incentives in each area may suggest the most attractive sites (with the best financial returns) to focus on for solar projects.

These incentives are frequently changing, so if you are located in an area with favorable incentives, you may need to move quickly to ensure you capture them. If you are located in an area without incentives, then consult with solar industry experts to learn whether new programs are expected to be implemented

WHAT IS A POWER PURCHASE AGREEMENT?

There are three primary parties involved in any PPA:

- **Host Customer**—The host customer is the building or land owner hosting the solar project. The project will be constructed on the customer's property, and must remain in place and in service for the term of the PPA (typically 15 to 25 years). The customer provides the roof or land area for the solar electric system, and will enter in to a separate (but required) sister agreement, called a Site Lease or Site License.

- **PPA Sponsor**—The private company that will own the solar project is the PPA sponsor (also referred to as the developer or PPA provider). Because the solar electric system itself is owned *not* by the building or land owner, but by a private third party, PPA structures are often referred to as "third-party ownership." The customer will enter a PPA directly with the sponsor. To understand the PPA financing, keep in mind that although the PPA sponsor is the entity dealing directly with the customer, the sponsor ultimately represents multiple entities, including:

  — Equity investors that will be able utilize the Investment Tax Credit (ITC) (also known as "tax equity investors"). These have historically been major banks and insurance companies, but other non-financial investors (such as Google) are moving in to tax equity investing as well.

  — Lenders providing debt capital to finance the project (not used in all cases).

  — Sponsor equity investors, who are responsible for managing the project.

  — A "special purpose entity" (SPE), usually a LLC, which will directly own the solar project. Each individual project is typically held in a LLC, to facilitate the financing and future sale of the asset.

  — In some cases, the sponsor may represent that it is constructing the project directly. In virtually all cases, the corporate entity constructing the project will be a contractor to the sponsor or the SPE that will develop, finance and own the project.

    It is standard and necessary in the industry for sponsors to represent the focal point for all of these entities, and sponsors are adept at managing these relationships.

- **Contractor**—The company performing the Engineering, Procurement and Construction ("EPC") work is referred to as the contractor (also known as the "integrator," "installer," or "EPC company"). Solar contractors will often be the host customer's first contact point, since contractors are regularly developing new business opportunities. Until the contractor meets with the customer and assesses its financing and ownership preferences, it will not be certain whether a PPA is required. If the customer elects to pursue a direct purchase (or an operating lease), it will contract directly with the contractor to build the system. If the customer enters a PPA, it will still interact closely with the contractor in analyzing the solar project, planning the schedule, access to the site, site evaluation, and ongoing project status updates. If the customer has developed a relationship with the contractor directly, the customer may require that the PPA specify the contractor that will build the project, to avoid any undesired changes over the course of the project.

Additional parties involved:

- **Utility**—The local utility may be involved in several possible ways, including interconnection, net metering, incentive allocation and payment, and other aspects of project approval. Many utilities in the United States offer solar incentives. The application and award process for these incentives varies. Some utilities have implemented standardized processes, while others award incentives through auction-like solicitation process, or impose other standards (such as the status of the project, or other discretionary approvals). The utility bears the ongoing responsibility for ensuring the solar project interfaces with the grid properly, and therefore must approve interconnection of the system. The timing and difficulty of this process varies based on the utility's technical requirements and grid quality at the site of each specific project. It is typically advantageous to consult with the utility before any sizeable project is initiated.

Figure 5-1 illustrates the contracting and financial relationships between these parties.

Under the PPA, the host customer agrees to purchase 100% of the solar electricity generated by the system, at a stated price. Electricity is denominated in units called **kilowatt-hours (kWh)**. From a financial perspective, the reason the customer would agree to pay for electricity generated by the solar system is that every solar kWh reduces the kWh

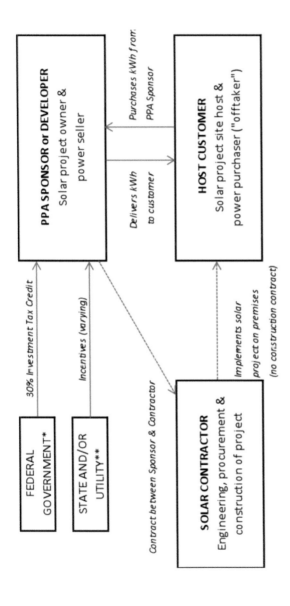

FEDERAL GOVERNMENT*

STATE AND/OR UTILITY**

30% Investment Tax Credit

Incentives (varying)

Contract between Sponsor & Contractor

PPA SPONSOR or DEVELOPER
Solar project owner & power seller

Delivers kWh to customer

Purchases kWh from PPA Sponsor

HOST CUSTOMER
Solar project site host & power purchaser ("offtaker")

Implements solar project on premises

(no construction contract)

SOLAR CONTRACTOR
Engineering, procurement & construction of project

*Note: For projects initiated or completed prior to December 31, 2011, the 1603 Cash Grant may be utilized in lieu of the Investment Tax Credit.

**Note: There are a range of incentive types provided by state and local governments, and by utilities. Certain states utilize private market-based incentives, such as Renewable Energy Certificates (RECs). Under market-based incentive structures, incentive payments may come from private entities.

Figure 5-1. Power Purchase Agreement—Indicative Structure

the customer must purchase from the utility.

A power purchase agreement may also be referred to as a Solar Services Agreement (SSA), or any number of variations used by individual PPA sponsors. Certain states and utilities restrict or prohibit PPAs. Depending on the nature of the restriction, PPA sponsors may develop alternative structures, such as leases, to synthesize some or all of the financial aspects of a PPA. In some cases, all third party ownership (which applies to both PPAs and operating leases, in which the lessor owns the system and takes the federal tax benefits) is prohibited. For each site being considered for solar, solar specialists will be able to advise you on the legality and financial viability of a PPA, or comparable structure.

The PPA sponsor will work with a solar contractor to design and build the solar project on the customer's property. The customer's first interaction will often be with the solar contractor's business development team. Solar contractors engage in sales and marketing efforts to develop new business. If the customer expresses objectives that suggest a PPA may be the best solution, then the contractor might team up with a PPA sponsor to finance the project. In some cases, the PPA sponsor may also do the design and construction directly, or through a related entity.

From the customer's standpoint, the important implication is that the host customer *does not enter a construction contract* when it enters a PPA. The customer agrees to host a solar electric system and purchase electricity generated by that system. The PPA sponsor with whom the customer contracts bears the responsibility to get the system constructed so that it can begin selling power. This distinction is noteworthy as you consider your objectives and concerns in pursuing a solar project. Any solar project is a sizeable physical structure that will remain on your property and occupy roof or land area for 20 or more years. Facilities managers will have reasonable concerns that the contractor and PPA sponsor will need to manage, such as building access during construction, roof access, roof warranties, safety, building and fire codes, and future access to the solar project site for maintenance purposes. From a contracting standpoint, however, the PPA is an agreement to pay for power or services, rather than a construction agreement.

In the PPA diagram, we highlight the government entities and the utility that may be providing financial incentives for the solar project.

Note that the arrows indicate these *incentives are going to the PPA sponsor, not to the customer.* In our experience, a common reaction from prospective solar customers is that a PPA would be an unfair scheme, as it would deprive the customer of the incentives that it would receive if it purchased the solar system directly. Of course, we cannot emphasize strongly enough that a *power purchase* agreement is extremely different from a *system purchase* or construction agreement.

The PPA sponsor and the host customer will only be able to come to an agreement if the customer is being offered a financially attractive rate at which to buy power (the "PPA rate"). The PPA sponsor, as we have seen, is responsible for handling system construction and financing it, as well as for capturing or "monetizing" all of the incentives. We will illustrate an indicative analysis of the PPA sponsor's costs and revenue streams in the text that follows. We focus on the flow of incentive funds (and tax credits) here to address the question "What is a power purchase agreement?" Under a PPA, the customer is doing one key thing: purchasing power. The PPA sponsor is investing in and paying for the system; negotiating with the contractor; procuring the federal, state and utility incentives; collecting payments from the customer for kWh delivered under the PPA; paying various taxes and permitting charges; and handling the operations and maintenance of the system.

To be clear, the customer is also providing the site, providing access to the site, managing any impact of the project on its daily operations, and so forth. What the customer does not need to do is manage the risk associated with the construction cost and timing, operations and maintenance of the system, or the responsibility of procuring the incentives. In fact, operational simplicity is one of the primary reasons solar customers cite for effectively "outsourcing" the solar project through a power purchase agreement.

Throughout the utility grid-connected areas of the United States, the cost of solar without incentives is currently high enough that any solar project without incentives (federal; and typically utility or state incentives as well) would require a PPA rate higher than the cost of grid power. Therefore, in order for a PPA sponsor to develop a solar project and offer the host customer a PPA rate less than the effective cost for grid power, it will need to take advantage of available incentives. The PPA sponsor will then "pass the savings on" to the host customer.

FINANCIAL DRIVERS OF PPAS (AND ALL SOLAR PROJECTS)

Solar energy projects and power purchase agreements are perceived to be complex undertakings—and certainly they require experienced solar contractors to install, and experienced financiers to develop PPAs. Our perspective, however, is that the fundamental financial drivers that enable successful, financially beneficial PPAs are straightforward to understand and quantify. It is not necessary to grasp every nuance of tax credit law or the photovoltaic effect to understand the basic P&L for a solar project. Moreover, we find that as our customers understand these financial building blocks, and how they enable the PPA to be financed, it enables the host customer and the PPA sponsor to work together creatively to anticipate hurdles, negotiate solutions and expedite projects. We acknowledge that the industry does not always make the financial equation transparent to the customer, and the purpose of this section is to remedy that.

These financial drivers are essentially the same for any project, whether the host customer purchases the system directly, or purchases power through a PPA rather than a turnkey system. This is not to say that the ultimate financial benefit is the same to the customer. Under a Purchase versus a PPA, various costs, responsibilities, and risks are borne by different parties, and each party will require a corresponding benefit. Our purpose here is to first explain the solar project "P&L," and then review the allocation of these cost and benefits under each structure; the five financial drivers include:

1. **Project Design and Price**
    The design of a project includes the type of mount (roof, ground, or parking structure), the technology (type of PV; fixed tilt or tracker) and development costs (permitting, environmental). These design factors are accounted for by the contractor in determining the Price of the project. All things being equal, the lower the price of the project, the lower the PPA rate will be. However, properly designed systems will optimize the technical, performance and financial factors to offer the best overall financial solution. A system utilizing higher efficiency modules or tracker systems will have a higher price, and will produce more kWh. Each project is unique and will require a different solution.

2. **Performance (kWh) and Value per kWh**

   The design of a project—specifically the location, orientation, solar panel tilt, and tracking technology (if any)—will determine the kWh generated by the project. The value to the host customer is the reduction in power purchased from the utility, and the associated reduction in the customer's utility bill. For the host customer, the kWh generated multiplied by the avoided utility cost per kWh determines its total utility savings.

   a. If the customer owns the project, this utility savings is the primary cash flow from the project.

   b. If the project is implemented through a PPA, then the PPA sponsor charges the customer a rate for every kWh generated. The customer's net cost or savings is the *difference between what it pays for the PPA and what it saves on its utility bill*. The revenue for the PPA sponsor is the product of kWh generated and the PPA rate.

3. **Government and Utility Incentives and Tax Benefits**

   a. <u>Federal</u>: Solar projects owned by commercial entities in the United States are eligible for a 30% federal investment tax credit (ITC) taken against income tax liability. The credit expires in 2016. Businesses may purchase systems directly and realize this credit. Government and non-profit entities are not eligible for the tax credit. However, commercial entities may be formed by PPA sponsors to own solar projects, realize the federal tax incentives, and sell power to customers at a negotiated rate (which in order to be competitive, essentially shares the financial advantages of the 30% ITC.) PPA sponsors may also sell power to commercial entities.

   b. <u>State</u>: State governments and utilities may offer solar incentives. These incentives vary significantly in structure and amount. Due to these variations, the same solar project at the same price may result in widely varying PPA rates in different states.

   c. <u>Tax</u>: Solar project capital costs are depreciable for income tax purposes. At the federal level, they may be depreciated over five years on the Modified Accelerated Cost Recovery Schedule (MACRS). State level depreciation rules vary.

4.   **Financing**

Financing costs for solar projects vary widely. PPAs require complex structures to enable private investors to monetize tax credits and depreciation benefits. The market to invest in tax credit projects is limited to companies that have taxable income, against which they can take the ITC. These financial structures require significant experience to implement, and the associated capital cost will vary from sponsor to sponsor. Capitalization structures for PPA vehicles are beyond the scope of this discussion. However, it is still relevant for facilities owners considering solar to understand the experience and track record of the PPA sponsor proposing the project. Despite the differences in capital structure and capital costs, the same P&L financial drivers will make the project more or less financially viable for any PPA sponsor. That is, a lower Price or higher kWh production figure will enable a lower PPA rate, regardless of the PPA sponsor.

5.   **Operating Expenses**

Solar photovoltaic projects are notable for the relative lack of operational requirements. Fixed-tilt solar PV systems have no moving parts, and do not require fuel. The solar panels are expected to produce power for more than 30 years (the oldest are now producing after more than 50 years!), and are warrantied by manufacturers to produce at a certain capacity (typically 80%) after 20 years or more. There are, however, operational requirements such as ~annual cleaning, as well as diagnostics on the electric components. The system must also be insured. After 15 to 20 years, it is likely that the inverter will need to be replaced. This is a larger cost, but still a small fraction of the total system value. Operating costs are typically small relative to the utility savings or PPA revenue. For single-axis and dual-axis trackers, there will be higher operating costs as there are moving pieces that require maintenance.

The financial drivers are summarized in Figure 5-2.

EXAMPLE SOLAR PROJECT FINANCIALS (SPF) ANALYSIS

This section is intended for non-technical and non-financial readers. However, because this is also intended as a practitioner's guide to managing a solar PPA project, it is important to carry the discussion from the

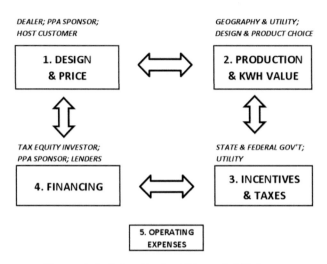

Figure 5-2. Solar Project Financial Drivers

general overview to a quantitative example.

First, we review quantitative assumptions, using realistic data points from recent projects. The reader should now have a sense of the factors that influence these assumptions, although the underlying details of how each figure is developed are not important for this example. Secondly, the Solar Project Financial (SPF) model P&L is presented in full detail, with calculations based on the following assumptions.

Assumptions

Our sample solar project P&L is representative of a project in the northeastern US.

- Project size: 500 kilowatts (kW). This represents a standard rooftop system on a large warehouse, box store, or ground area.
- Project price: $3.75 per W; totaling $1,875,000.
- Solar energy production: 625,000 kWh per year. This is expressed as kWh production on a per-Watt basis, at 1,250 kWh per kW per year. Assumed to decline at 0.5% per year, which is known as "degradation."
- PPA Rate: $.12/kWh in Year 1.
- PPA Escalation Rate: 3.0% per year.
- Incentive: $.15 per kWh ($150 per MWh) for the 15 years. (This assumes an incentive that is paid on a kWh basis, at a fixed rate for a set period of time. Not all incentives are based on kWh generation, but this is the predominant incentive structure nationally.)

- Operations and maintenance (O&M): $.015 per Watt per year. Inverter replacement in Year 15 at $.25/W.

- Federal tax rate of the PPA equity investors: 35%. No state tax rate is assumed. (Even if state tax benefits are realized, the impact is often too uncertain to impact PPA pricing.)

- Internal rate of return (IRR) calculated on an after-tax basis, incorporating the revenues, expenses and tax impacts listed above. (See Figure 5-3.)

Figure 5-4 is a profit and loss (P&L) example numbered and color coded to correspond with the preceding financial drivers diagram.

PROJECT SIZE & PRODUCTION:

PROJECT SIZE:	**500.0**	kW-DC
Production factor:	**1,250**	kWh / kW / year
Annual solar production:	625,000	kWh / year

PPA REVENUE

PPA Rate:	$ **0.120**	/kWh
Annual PPA Revenue:	$ 75,000	year 1
Rate Escalation:	**3.0%**	/year

INCENTIVE REVENUE

PBI/SREC/FIT:	$ **0.150**	/kWh
Term of PBI/SREC/FIT:	15	years

EXPENSES

Operations & Maintenance	$	0.015 $ / W / Year
Inflation		3.0%
Inverter replacement cost	$	0.25 $ / W
Inv replacement in year:		15

TAXES

Fed Tax:	35%
State Tax:	0%

Depreciation

Basis		Federal
Total Price		$1,875,000
Less: 50% of ITC	15%	($281,250)
Depreciable Basis		$1,593,750

Fed Depreciation Schedule*	5 yr DDB	20.00%	32.00%	19.20%	11.52%	11.52%	5.76%

*MACRS depreciation schdule - indicative assumptions.

Figure 5-3

EXAMPLE: SOLAR PROJECT PROFIT & LOSS (P&L) STATEMENT

Year	0	1	5	10	15	20	
Production - kWh		625,000	612,500	596,875	581,250	565,625	PRODUCTION (2)
Total Price	$3.75	$(1,875,000)					PRICE (1)
ITC	562,500	-					INCENTIVES (3)
State Tax Credit	-						
Cash Flow							
PPA Rate:	$ 0.120 /kWh	$ 0.120	$ 0.135	$ 0.157	$ 0.182	$ 0.210	VALUE OF
Energy Sale Revenue		$ 75,000	$ 82,725	$ 93,454	$ 105,503	$ 119,019	PRODUCTION (2)
Incentive Revenue							
PBI/SREC/FIT Revenue	$ 0.150 /kWh	93,750	91,875	89,531	87,188	-	INCENTIVES (3)
O&M Expenses		(7,500)	(8,441)	(9,786)	(11,344)	(13,151)	OPERATING
Inverter Replacement		-	-	-	(125,000)	-	EXPENSES (5)
Pre-Tax Cash Flow	(1,312,500)	161,250	166,159	173,200	56,346	105,868	
Fed Tax		55,125	6,104	(60,620)	(19,721)	(37,054)	TAXES (3)
State Tax		-	-	-	-	-	
Cash Flow	($1,312,500)	$216,375	$172,263	$112,580	$36,625	$68,814	
IRR				5.2%	8.8%	9.9%	FINANCING (4)
Cumulative Cash Flow	$ (1,312,500)	$ (1,096,125)	$ (254,902)	$ 330,701	$ 827,935	$ 1,156,451	

Figure 5-4. Profit and Loss Example

Summary and PPA Financing Implications

Based on the assumptions listed above, the solar project P&L can be estimated over 20 years. The revenues, expenses and tax benefits are added each year in the line labeled "Cash Flow." This represents cash flow after tax benefits. Internal rate of return (IRR) is a common financial metric used to measure the time-adjusted returns on an initial investment.

The 20-year after tax IRR in the example shown is 9.9%. The IRR line is labeled "Financing (4)." As noted in the discussion of financial drivers, every PPA sponsor has its own financing criteria. Sponsors typically have a target investment return, which will vary by sponsor, and will change over time as market conditions fluctuate. As of this writing, an IRR in the 9% to 10% range is consistent with what we are seeing in successful PPA transactions for creditworthy companies and municipal agencies.

In practice, PPA and financing practitioners each refer to investment returns in different, highly specific ways. Returns may be calculated before or after taxes, and over different numbers of years. For example, in the P&L below, the 10-year IRR is only 5.2%, and the 20-year IRR is 9.9% (after-tax). This is not inconsistent; in fact, it highlights the core investment thesis, which is that solar energy is a long-term financial and energy generation solution. IRRs may also be expressed on a "leveraged" basis, which applies to projects that are utilizing debt. For all projects, the metrics used to evaluate returns to the tax equity investors will differ from those used by the sponsor equity investors.

These financial returns are the proprietary business of PPA sponsors. We mention them here to highlight that host customers do not need to calculate them, nor seek full financial disclosure from PPA sponsors. Customers should focus on understanding the high level drivers that we have listed, and target their efforts around developing the most economic project possible.

The Customer's Perspective

The next question is whether the PPA financial terms outlined above—an $.12/kWh PPA with a 3% escalator—would be compelling to the host customer.

Each customer will have its own financial concerns and criteria. Over 10 years of developing solar projects from Hawaii to New Jersey, we have seen customers with a range of financial requirements. Certain customers have experienced very high utility rate inflation or rate volatility. This situation has been acute in Hawaii and California over the past decade, for example. These customers may seek to lock in a predictable rate for electricity over 20 years, and may be willing to pay a premium to current prices to achieve that certainty. These customers are looking to hedge their exposure to utility rates, just as many companies hedge exposure to commodity prices or interest rates.

Many companies and municipal agencies, experiencing difficult financial situations, are focused on moderate immediate reductions in their utility expenses, with the long term potential for huge savings once the system is paid off, or the PPA expires, and they buy out the system. These customers require PPA rates (and rate escalators) that are lower than the per-kWh savings they expect on their utility bills (coupled with their expected annual utility inflation rates).

For example, we extend the PPA example above to show the financial benefit to a host customer that is able to achieve utility savings of $.13 per kWh. This is indicative of the typical cost of power in northeastern US markets. See Figure 5-5.

In this example, the host customer realizes immediate financial savings by purchasing power from the PPA sponsor at a lower rate than it is paying the electric utility. Equally importantly, the customer is locking in a predetermined rate for its power over 20 years. The assumed savings in this model is based on a utility inflation rate of only 3% per year. However, over the past 40 years, utility rates nationally have increased at an average of over 5% per year. In certain states, utility inflation has been more pronounced than elsewhere.

The host customer in this example has achieved its objective of implementing a solar energy project to generate electricity from sustainable sources, while achieving financial savings from day one, and hedging its utility rate exposure for 20 years or more. Results will vary for every specific project. We hope this financial analysis supplies a conceptual model for understanding the requirements of the financial investors, as well as the benefits to the host customer.

ADDITIONAL MATERIAL—
FOR FUTURE EXPANSION OF THIS SECTION

Project Design and Price

As with any capital investment decision, the primary issue is how much it will cost to implement. We refer to this as the "Price" (sometimes

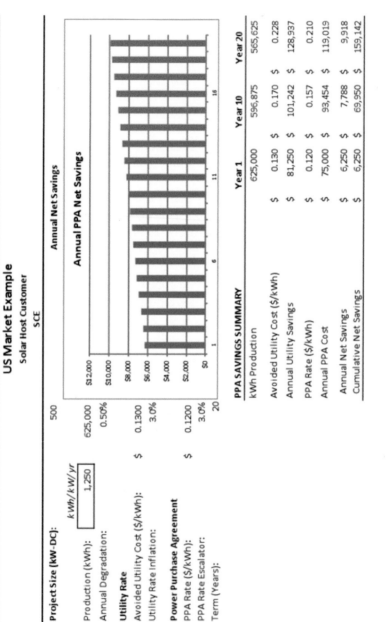

Figure 5-5. Solar Project Financial (SPF) Model
Power Purchase Agreement (PPA) Savings Analysis

also referred to as the "Turnkey Price" or "EPC Price"). There are innumerable design and engineering considerations that go in to determining the price of a solar project. Equally importantly, there are non-technical considerations that impact and define the design requirements: aesthetic, environmental, financial, operational, and so on. The critical point we highlight is that these myriad considerations will impact the design of the project, the price of the project, *and consequently, the price of solar power purchased under a PPA.*

Power purchase agreements are often advertised and implemented as a "no money down" solar solution for host customers. This results in solar projects and PPAs being viewed as a black box, with no connection between the inputs (such as the design and price of the project) and the outputs (namely, the price per kWh generated). The unfortunate irony is that, for customers requiring a "no money down" solar solution, they often put a significant amount of resources (time, if not actual cash) in to a detailed set of technical, operational and other specifications. Facilities managers, engineers, consultants, and even lawyers meticulously detail each of their requirements, on the assumption they can throw them in to the black box, and generate a no money down PPA on the back end. The job of explaining the specific impact of a given requirement rests with the contractors and PPA sponsors competing for the project—and each project will have its own unique challenges that competitors will address with different solutions. We hope that awareness of this relationship between requirements and costs will facilitate more efficient evaluation of potential projects.

As a basic overview, there are three types of projects, based on the nature of the site. Each of these project types will have advantages and disadvantages.

Roof Mount

Roof mounted solar is the most common solution for government and commercial buildings, because it is often cost-effective to install, and utilizes roof area with limited alternative use. Roof requirements will vary, however. Any solar project, particularly a PPA project, will need to remain in place uninterrupted for 20 or more years. Therefore, the roof must be relatively new, so that it will not require replacement during this timeframe. It is prohibitively expensive to remove a solar project, and then reinstall it. The labor expense and lost energy production make this a non-starter. For building owners that will not accept penetration of the roofing membrane, solutions such as ballasted systems may be feasible. The size, obstructions, and orientation of the roof will impact the tilt and orientation of the solar panels installed. Solar panels can be installed with no tilt, but energy production will be higher when panels can be tilted to face southward. The contractor and PPA sponsor will work together to address these issues. The host customer (building owner) should understand that competitors will develop different solutions to optimize the price and performance of the system, and that will impact the PPA rate offered.

Ground Mount

For those customers with available land, a ground mounted solar project has the potential to be more cost-effective than a roof mounted project. Ground area may offer easier access, with more options for adjusting tilt and orientation to maximize solar energy output. However, ground areas may also be subject to stricter permitting and environmental requirements. Civil engineering and grading costs may be considerable, and subsurface soil conditions may create uncertainties and surprises that can drive up construction costs. Landowners may have alternate uses for the land, that would factor in as either soft costs (precluding expansion of a business facility) or hard costs (direct loss of rental revenue). Any opportunity costs would need to be offset by the benefits of the solar project to the host. Finally, we would highlight that the land area available for solar would need to be reasonably close to the utility meter (or other interconnection point at the facility), or added transmission costs would be required.

Elevated Parking Structure

Parking lots and pedestrian areas may offer potential for solar through the use of elevated parking structures (also referred to as canopies, elevated racking structures, and various other terms). Elevated parking structures offer a creative way to take advantage of such spaces, and have been implemented in many projects nationwide. These structures must be stable to withstand wind and seismic requirements, which requires a significant amount of steel. As of this writing, steel prices are increasing and parking structures often result in a cost premium to roof mounts. Conversely, when roof area is limited or unavailable, parking structures may offer a financially viable alternative.

Chapter 6

PACE Financing

Anthony J. Buonicore, P.E., BCEE, QEP

INTRODUCTION

The commercial real estate (CRE) market in the U.S., consisting of approximately 4.8 million office, retail, service, lodging, multifamily, warehouse and storage buildings, represents a significant opportunity for building owners to reduce energy use and monetize their energy savings. Moreover, it is now evident to CRE owners and lenders that building energy performance can impact property value. As a result, less energy efficient buildings are at a growing competitive disadvantage and in danger of accelerated obsolescence. In property transaction due diligence, for example, knowledgeable buyers now consider subpar building energy performance no different than any other property deficiency such as a damaged roof or an air conditioning system at the end of its useful life.[1]

These market developments have stimulated a growing number of retrofit projects designed to increase energy efficiency. To the extent that energy efficiency investment has been made in the CRE market, it is most likely associated with lower cost improvements having relatively short payback periods (less than 2-3 years) and involving low technology risk. As a result, the CRE industry now has the opportunity to move from this initial phase of low cost, short payback energy efficiency improvements to the multifaceted second phase of implementing deep energy retrofits (defined as resulting in at least a 30% reduction in whole building energy use) where the capital need is much more intensive and the payback period often longer.

In view of scarce internal funding and the desire to preserve capital in an uncertain economy, CRE owners have taken a measured approach toward the opportunity to monetize potential energy savings. There are two principal reasons for this pragmatic approach to capturing these savings. The first is associated with the current behavior of the CRE market itself and the second is associated with the availability of commercially-attractive financing.

Over the last four years, commercial real estate has been a victim of the country's most severe recession since the Great Depression and the accompanying economic uncertainty. Vacancy rates escalated and rent growth has been virtually non-existent. As late as last summer when the CRE market appeared to be gaining some traction, along comes an onslaught of more disappointing economic news and job growth setbacks, including downward revision of the gross domestic product (GDP), fiscal chaos in Europe, debt ceiling gridlock, a downgrade of U.S. debt, and significant volatility in the investment market. A double dip recession seemed just over the horizon. These economic headwinds resulted in further CRE market stagnation and the modus operandi again became "preserve capital to the maximum extent possible." Capital expenditure (CapEx) budgets plummeted, operating budgets were meticulously scrutinized and equipment replacement or upgrades were often put on hold. Where possible, even maintenance was being deferred.

To make matters worse, given that almost 75% of CRE was constructed prior to 1990, and that many of these buildings still rely on original mechanical and electrical equipment often near the end of its useful life, this has resulted in substantial pent-up demand for equipment upgrades and replacement.

Fortunately, to-date during 2012 the economic climate has shown signs of improvement and the country may now hopefully be on the road to a sustained recovery. Assuming this to be true, the floodgates holding back the substantial pent-up demand for equipment replacement and upgrading may finally be at the cusp of opening. This dynamic will represent a significant opportunity for replacing or upgrading dated energy-consuming equipment with much more efficient units. The end-result of this powerful business driving force will likely be rapid acceleration of the deep energy efficiency retrofit market.

The execution challenge associated with these deeper, more capital-intensive energy efficiency retrofit improvements is complicated when internal financing is limited or not available. While some financing for energy efficiency upgrades has been available to CRE owners, the availability of "commercially-attractive" financing often has not. Fortunately, this is changing and market ready, commercially-attractive financing mechanisms have arisen to meet the need.

This chapter will review these market ready, commercially-attractive financing mechanisms and the emerging best practice needed to facilitate proper underwriting of energy efficiency loans. The net result will be energy efficiency lending finally becoming a mainstream financial asset class with a high degree of standardization, predictability and scale.

> *"The flood gates holding back the substantial pent-up demand for equipment replacement and upgrading may finally be opening...the end-result of this powerful business driving force will likely be rapid acceleration of the deep energy efficiency retrofit market."*

KEYS TO ENERGY EFFICIENCY INVESTMENT

There are four requisites for CRE building owners contemplating external financing for an energy efficiency investment. First, such external financing must be easily accessible and available with "commercially attractive" terms. Secondly, the investment must be based on a reliable and fully transparent methodology to project future energy savings with a high degree of confidence. Thirdly, actual energy savings performance after improvements are made must be measurable and verifiable in a reliable, consistent and fully transparent manner. Lastly, the risk of underperformance must be low. The first and fourth requisites focus more on the financing structure, while the remaining two requisites focus more on the technical underwriting.

> *"While some financing for energy efficiency upgrades has been available to CRE owners, the availability of commercially-attractive financing often has not. However, there now are a number of financing mechanisms that can meet the commercially attractive financing criteria."*

COMMERCIALLY ATTRACTIVE FINANCING

While financing is commonly available from multiple sources to support energy efficiency investment, finding "commercially-attractive" terms has often been problematic. "Commercially-attractive" terms can mean many things, but for the purposes of this chapter it will be "ideally" defined as financing:

- without any capital expense;
- that does not add debt to the property;
- that covers 100% of the project cost, including all upfront [hard and soft] costs, such that there is no "out-of-pocket" owner expense;
- structured such that payments can be treated as an operating expense;
- structured such that payments (along with the energy savings) can be passed along to tenants (in a multi-tenant building); and
- available at relatively low cost (interest rate) and payable over an extended period of time (10 years or longer), such that monthly energy savings can more than offset the monthly payment necessary to capture these savings, thereby enabling projects to achieve cash flow positive status immediately.

There are a number of financing mechanisms that can meet the "commercially-attractive" financing criteria. These can be extracted from Table 6-1(3) and include:

- PACE tax-lien financing
- Energy Service Company (ESCO) direct financing
- ESCO third party financing using the PACE structure
- Energy service agreement providers using private party financing
- Energy service agreement providers using PACE financing
- Bank debt provided through a PACE structure

Property Assessed Clean Energy (PACE) Tax-Lien Financing

PACE tax-lien financing programs allow local governments, when authorized by state law, to fund energy improvements on commercial and industrial properties via an additional assessment on the property tax bill. Similar to a sewer tax assessment, loans under a PACE program, are secured by a lien on the owner's property and re-paid through an assessment on the owner's property tax bill. This structure results in a lower cost of capital payable over a long term (typically 10-20 years). PACE financing transfers with sale of the building so that future owners or tenants assume the payments, along with the continued cash flow positive energy savings benefit. Mortgage holder's consent is often required in many states before applications can be approved and assessments placed.

Early stage commercial PACE programs in Sonoma County (CA) and Boulder County (CO) funded projects with, in the case of Sonoma, ex-

Table 6-1. Comparison of Financial Options

EE Retrofit Financing Options	No Capital Expense	No Upfront Investment for Audit	100% Project Financing	No Debt on Property	100% Write-off of Annual Payments	Lower Cost of Capital	Longer Term Financing	Cash Flow Positive
Traditional								
Internal			N/A	✓	*	N/A	N/A	N/A
Bank Debt					*			?
thru/PACE**	✓		✓	✓	✓	✓	✓	✓
Lease	✓		✓	✓	?		?	?
Non-Traditional								
ESCO	✓	✓	✓	?	?	?	?	?
w/PACE	✓	✓	✓	✓	✓	✓	✓	✓
ESA	✓	✓	✓	✓	✓	?	?	✓
w/PACE	✓	✓	✓	✓	✓	✓	✓	✓
Government Loan						✓		?
PACE	✓		✓	✓	✓	✓	✓	✓
On-bill Utility	✓		✓	✓	✓	✓		?

* Accelerated depreciation of capital investment

** "Owner-arranged financing" PACE Model

✓ indicates this funding criterion can be met

? indicates this funding criterion may or may not be met depending on the specifics, e.g., an operating lease expense can be written-off each year, while a capital lease expense may not

A blank space indicates that this funding criterion is not met.

isting county treasury funds, and, in the case of Boulder, municipal bonds of the county. Funding from existing reserves is an appealing option for a number of reasons: funds are available when projects need them, and an interest rate can be applied to the project that is attractive to both the property owner and the government doing the lending. Unlike Sonoma County, however, most local governments are unlikely to have substantial reserves from which to lend. Boulder County, on the other hand, established a pool of projects, and when there was an aggregate demand that could support the efficient sale of bonds, all projects were funded simultaneously. However, without a fairly steady stream of projects, building owners will likely have to wait longer than they would prefer to have projects funded.

Other funding models are being explored by emerging PACE programs. One, a "warehouse" model involves an investor (such as a large money center bank) providing a line of credit for the cities and counties to use in funding the PACE program. In such cases, the bank envisions warehousing the loans until a critical mass is reached at which time bonds or other securities can be issued in order to replenish the line of credit. As part of the process, the warehouse lender needs to include the cost associated with hedging interest rate risk before critical mass is reached. As such, good project flow is crucial in that it allows critical mass to be reached quickly and will minimize hedging costs.

The second is the "bond" PACE model that involves the issuance of bonds to create a local or state fund that the local government will then make available to the PACE program. Once the bonds are sold and the PACE program funded, the "bond" model is similar to the "warehouse" model.

Recently launched PACE programs in San Francisco and Los Angeles are using what is being referred to as an "open market" model (or also referred to as a "private placement" or "owner arranged" model) where financing is provided by private investors, which could be banks or pools of funding raised from private investors. This is expected to be a very attractive model in the CRE market. The municipality acts as a conduit for private investment. Individual property owners arrange their own financing directly with the project lender leveraging the enforceability of the tax lien on the property as security. This enables building owners to negotiate rates, terms, conditions, and schedules that best suit their specific project needs, rather than waiting to lock in a rate through a bond. The owner-negotiated terms are then reflected in a loan agreement directly with the lender. Financing is repaid as a line item on the owner's property tax bill. The repayment obligation transfers with ownership. This PACE model opens a wider channel of capital inflow compared to pooled bond models.

In the CRE industry, where there are a significant number of triple-net-lease tenant-occupied properties, a significant advantage to these PACE assessments is that they normally qualify as operating expenses under existing leases and, therefore, are eligible "expense pass-throughs" to tenants. Under typical triple-net lease agreements where tenants are responsible for utility costs, the pass-through of the PACE assessment as a tax reimbursement allows owners to implement projects and equitably share project costs with the tenants who in return reap the benefit of lower energy cost.

The security of the tax lien also provides a solution to the inability of many commercial building owners, who often lack investment-grade credit ratings, to secure any type of third party financing for energy retrofits. The lien is attached to the property and transfers with ownership. Repayment security is through the senior lien position of the assessment rather than through the borrower's credit. This allows owners to undertake deeper retrofits with greater energy savings and longer payback periods, even if the owner only plans to hold the property for a few years.

The process of owner-arranged PACE financing begins when the building owner engages an energy service company (ESCO) to audit the property and develop a retrofit plan. The owner then submits the plan to the municipality for approval, in some cases along with a lien consent letter from the mortgagee. Once the municipality notifies the owner of approval, the owner can negotiate financing from lenders on advantageous terms due to the security of the lien, which will be placed on the property when funding is provided. The owner will then typically enter into an energy savings performance contract with the ESCO, and the lender pays the ESCO to perform the installation. The municipality assigns the assessment collection rights to the lender, and the building owner pays the assessment according to the agreed upon schedule. The ESCO provides operation and maintenance (O&M) and energy savings measurement and verification (M&V) for a service fee and pays the owner if verified savings fall short of the energy savings guarantee.

Typical PACE tax-lien financing structures make it possible to have the reduced monthly energy bill (reflecting the energy savings) more than offset the additional charge (for loan repayment) on the monthly property tax bill enabling immediate positive cash flow. To date, 28 states and the District of Columbia have passed enabling legislation enacting PACE programs. More than a dozen commercial PACE programs are actually in operation or are well along in the development process.

PACE Advantages for the CRE Industry

The CRE industry will find PACE programs attractive for a number of reasons.

1.   For multi-tenant, investment property, costs (and associated savings) can be passed to tenants under existing leases.
2.   The loan is secured by the tax lien on the property rather than the borrower's credit.
3.   Building turnover is irrelevant since PACE financing transfers with the sale of the building and future owners assume the payments (and benefit from the savings).
4.   100% of the project cost can be financed.
5.   There is no additional debt on the building.
6.   The PACE structure will result in positive cash flow immediately since the savings will more than offset the costs.

Underwriting Criteria for PACE Financing

In the final analysis, whether a project can utilize PACE financing depends on the property owner's ability to pay assessments as evidenced by the financial strength of the project. The "ideal" PACE project satisfies the following underwriting criteria:

1.   The project should involve high-value improvements involving significant energy efficiency gains.

2.   The value of the real estate relative to project financing -the ultimate security for an assessment-backed obligation
      — needs to be carefully examined to determine whether an energy efficiency project is financially viable. Existing debt on the property, together with the PACE assessment obligation, should be significantly less than the value of the property. If the mortgage and other debts on the property exceed the property value, there is an increased risk of default and such projects will not qualify. There is a preference that the existing loan-to-value ratio associated with the property should not exceed

85% before improvements. There is also a preference that the maximum lien-to-property value ratio be 15% to ensure that any delinquent, uncured PACE assessment that is payable senior to the mortgage upon default is nominal in value compared to the outstanding mortgage.

3. The property should have clear title with no encumbrances. Property taxes should be current. There should be no recent bankruptcies, no outstanding liens on the property, or notices of default or evidence of debt delinquency. The property owner should be current on mortgage payments and there should be no easements or subordination agreements that would conflict with the PACE assessment.

4. The project should pay for itself, i.e., the projected monthly energy savings should be greater than the expected monthly cost of the PACE assessment over the term of the PACE loan.

5. The project should have a useful life longer than the term of the projected financing.

6. Credit enhancements such as availability of a state loan loss reserve fund, or a letter of credit, or the use of energy savings insurance, or the availability of federal or state loan guarantees, will make the credit profile more attractive and enable more attractive financing terms.

ESCO Financing

An ESCO represents a one-stop shop for project development and installation. Many large ESCOs with significant financial resources (such as Johnson Controls, Honeywell, Siemens, Eaton, Schneider Electric, Chevron Energy, Trane, Ameresco, etc.) also provide project financing. Projects are typically large-scale with the contract period covering a 5-10 year period or longer.

Various types of energy savings performance contracts (ESPCs) exist, including "shared savings" contracts, "paid from savings" contracts, and "guaranteed savings" contracts. Under typical ESCO contracts, newly installed equipment is financed, owned and maintained by the ESCO. Ownership transfers to the building owner at the end of the ESPC period. It may be accomplished by either a purchase at fair market value or the building owner may simply assume ownership of the equipment that has been paid for during the ESPC term. The majority of ESPCs are financed through savings generated by reduced energy consumption.

With the "shared savings" contract, the dollar value of the measured energy savings is divided between the building owner and ESCO. If no energy cost savings are realized, the owner continues to pay the energy bill, but does not incur any expense to the ESCO for that period. In the "paid from savings" contracts, the building owner pays the ESCO a predetermined amount each period (for example, an amount equal to 80% of the expected energy bill had the improvements not been made). Under "guaranteed savings" contracts, the ESCO guarantees that energy cost savings will exceed an agreed upon minimum dollar value. To ensure a positive cash flow to the owner during the ESPC term, the guaranteed minimum savings typically equals the financing payment for the same period. ESCO pricing often includes a fee that covers on-going monitoring, measurement and verification costs and a premium for assuming underperformance risk.

To date, the majority of ESCO work has been performed in the Municipal, University, Schools and Hospital (MUSH) market, principally because the ESCO business model is based on large, long-term ESPC contracts and significant government funding is available. It requires clients like MUSH owners who typically have very large energy efficiency retrofit projects (for example, involving multiple buildings on a university campus) and are committed to operate their properties for relatively long time spans.

ESCO Financing Using the PACE Structure

ESCOs generally either provide their own financing or bring in a third party financing source. Operating under a PACE structure, however, allows the ESCOs to offer their services to a project that had obtained "commercially-attractive" financing. Thus, for ESCOs who prefer not to provide their own financing, the availability of a PACE funding structure eliminates the need to locate interested third party funding sources. The lender(s) already in the PACE program would simply pay the ECSO upon completion of installation and verification of the energy savings. The ESCO would, of course, still be at risk if the verified savings fall short of the energy savings guarantee.

While ESCOs have made some progress in the owner-occupied segment of the CRE industry, this has not been the case in the multiple tenant segment (or the traditional CRE investment sector), where building turnover is much more frequent and often opportunistic, i.e., on average every 4–7 years. However, use of the PACE financing structure would allow ESCOs to expand into the much larger multi-tenant building sector. With the lien attached to the property and not the property owner, ESCOs can undertake in both of these CRE sectors deeper retrofits with greater energy savings and longer payback periods, even if the owner only plans to hold the property for a few years. This will result in a significantly broader target market for ESCOs.

Energy Service Agreements

A number of innovative managed energy services agreement (ESA) structures are now being offered by third parties who develop projects, arrange or provide the capital, and manage the installed equipment. These typically are pay-for-performance solutions where energy efficiency is essentially being sold as a service. Energy efficiency service providers are compensated only if energy savings are realized. Building owners have no upfront cost, no capital requirement, and 100% of the project cost is financed. The ESA provider assumes ownership and maintenance responsibility for project assets over the lifetime of the project. Payments to the energy efficiency service provider are viewed as a "pass-through" operating expense (to building tenants).

There are a growing number of energy efficiency service firms offering pay-for-performance financing solutions under ESAs, including SCIenergy/Transcend Equity Development (founded in 2002, Dallas, TX), Metrus Energy (founded in 2009, San Francisco, CA) and GreenCity Finance (founded in 1990, Indianapolis, IN).

Under the Transcend model, building owners pay Transcend a service fee based on historical energy costs. Transcend, in turn, pays the utility bill and earns its fee from savings generated by the efficiency improvements. The Transcend fee becomes an operating expense (pass-through to tenants) that replaces the utility bill and the building owner incurs no debt. At the end of the ESA term (typically 5-10 years), title associated with the improvements passes to the owner. If the building is sold, the contract can be assigned to the new owner (or terminated if preferred). Transcend will typically enter contracts where they envision at least a 25% savings on the current utility bill. The company's ideal customer has a minimum aggregate space of 250,000 square feet, associated with one or more buildings. This is a relatively large building or complex.

Under the Metrus model, in contrast, building owners maintain responsibility for payment of their reduced utility bills (which directly benefits tenants in a multi-tenant property) and pay Metrus's fee (which is a pass-through operating expense paid by the tenant) out of the delivered energy savings. The Metrus fee is structured as a per-unit-saved payment (i.e., a price per avoided kilowatt hour of electricity and/or avoided therm of natural gas), where the price for energy unit savings is set at a level below the prevailing utility price per unit of energy consumption. This arrangement establishes energy efficiency as a resource and is akin to a solar power purchase agreement, where the customer has no project performance or technology risk and pays only for realized, measured and verified energy savings. Metrus retains ownership of all project-related assets for the duration of the ESA term. At the end of the contract period, clients can purchase the equipment for fair market value. Metrus works with ESCO partners and typically prefers clients to have approximately $1 million or more in combined electricity and natural gas costs annually. Their energy efficiency projects typically have a payback period in the 3-7 year range. Metrus' business model also can include energy savings insurance. To-date, Metrus has focused principally on owner-occupied buildings.

GreenCity Financing provides a proprietary off-balance-sheet financing model that shares energy savings with the building owner. The investment is maintained as an operating expense and paid for out of the energy savings. This model assumes no out-of-pocket costs to the building owner and the risk of performance failure is assumed by GreenCity.

Energy Services Agreements using the PACE Structure

ESA providers typically incorporate third party financing through relationships with multiple financing sources. Operation within a PACE structure would likely bring more commercially-attractive financing terms (longer duration loans at more attractive rates) and eliminate any need to locate interested third party funding sources. Moreover, use of the PACE financing structure would allow an ESA provider to pursue deeper energy retrofits within a much larger market, i.e., the many commercial buildings where owners lack investment-grade credit ratings. With the lien attached to the property and not the property owner, ESA providers can undertake deeper retrofits with greater energy savings and longer payback periods, even if the owner only plans to hold the property for a few years. This will result in a significantly broader target market for ESA providers.

There are a number of reasons why lenders are seriously considering the benefits that will accrue to them by participating in PACE programs."

Bank Financing within PACE Programs

Under the "open market" PACE model, financing can be provided by private investors, such as banks, who have traditionally provided debt financing. The problem with traditional debt financing has been that it has been relatively expensive, highly dependent on the borrower's creditworthiness, unable to fund 100% of a project's total cost, and rarely available for much longer than approximately 5 years. However, when working within a PACE program, reliance shifts from the borrower to the property. Moreover, the nature of tax lien financing can result in a credit enhancement that reduces risk and therefore should lower the cost of capital.

Consent of the Existing Mortgage Lender

Since the lien associated with a PACE loan occupies a priority position, the mortgage holder's consent is typically required before PACE applications can be approved and the assessment placed. Commercial mortgages almost always give an existing lender the right to approve an additional senior or subordinate debt, and even voluntary tax assessments in some cases. Lenders also typically have the right to approve any structural changes to a building or its operating equipment since the building in its entirety represents the lender's collateral.

Much has been said in the development of PACE programs about potential bank resistance to the priority position of the PACE loan. However, a broad range of commercial PACE projects have already received consent thus far from a mix of national, regional, and local mortgage lenders. Moreover, there are a number of reasons why lenders benefit by participating in PACE programs.

1. With existing building owner customers who own property that can readily be made more energy efficient, it is an opportunity for banks to increase business with these customers. Building owners find the increased cash flow, the accompanying increase in building valuation, and the improved competitive position the building has in the market very attractive. For the bank, it is also an opportunity to identify and solicit new customers from other institutions that do not participate in the PACE program.

2.   Much of the energy efficiency work will likely be associated with replacing energy-consuming systems that have exceeded their useful life, work that has been delayed principally due to the recession. By providing loans to these existing customers, banks will be protecting their collateral and helping the property avoid obsolescence.

3.   In case of default, non-acceleration clauses associated with typical PACE programs require that only the low monthly payments be paid by the foreclosing bank, a fact that significantly reduces the financial impact on the lender of the PACE's priority lien in the event of a default.

4.   A secondary market (securitization) for PACE loans is already being discussed. The ability to package loans and sell them on the secondary market would be attractive to banks who could then re-lend the replenished capital as it sees fit, a process that can positively impact a bank's profitability.

5.   Federal energy efficiency loan guarantees and/or state energy efficiency loan loss reserves (that may be associated with the PACE program) can provide credit enhancement and reduce default concern.

6.   Emerging energy savings insurance, which is able to guarantee the energy savings (from which the lender is recovering both capital and interest), can also provide a credit enhancement.

7.   PACE programs can support a bank's commitment to sustainability, creating an opportunity for excellent public relations in the community and within the customer base.

To date, lenders have had a difficult time getting their hands around energy savings because energy savings cannot be measured directly. Energy savings are based on what is not going to happen in the future, rather than what will happen. Moreover, cash flow from future energy savings is not a familiar form of revenue or collateral that has been used to secure bank lending. There has also been a general lack of confidence in energy savings projections because of the embedded bias to present projects as compelling investment opportunities.

Another challenge lenders have faced is associated with the relationship between a building's energy performance and its value. To-date, there is insufficient data on how building valuation is impacted by energy efficiency improvements. Appraisers have not focused on a property's energy efficiency and therefore it is not reflected in their valuation. This void creates uncertainty and adds to the potential risk associated with energy efficiency investment.

Notwithstanding, lenders are beginning to recognize that energy efficiency loans can help preserve the value of an existing customer's building by avoiding obsolescence. In fact, the obsolescence issue, directly related to the value of the collateral, is an important consideration to lenders. It is something they understand, and may even be a more important consideration today than operational savings.

Furthermore, the emergence of standardized measurement and performance protocols that can successfully identify energy savings with a high degree of confidence(4) are poised to provide lenders with a clear understanding of the proposed benefits or lack thereof. By building these new protocols into the loan documentation and the underwriting process, lenders are becoming more comfortable with the way energy savings and risks can now be quantified. In the final analysis, a better understanding by lenders of energy efficiency investment, along with recent developments and tools to improve underwriting, will enable energy efficiency financing to become a mainstream financial asset class with a high degree of standardization, predictability and scale. Combined with the positive reinforcement this can provide to rating agencies and investors, it should also go a long way toward moving the CRE industry to large scale adoption of energy efficiency investment.

"Lenders have had a difficult time getting their hands around energy savings because energy savings cannot be measured directly. Energy savings are based on what is not going to happen in the future, rather than what will happen."

CREDIT ENHANCEMENTS

Default Protection

President Obama's "Better Building Initiative," announced in February 2011, calls for a federal loan guarantee program (run through U.S. DOE) to encourage private lenders to embrace energy efficiency retrofit financing. This proposed government guarantee for qualified energy efficiency loans is a contractual obligation between the government, private creditors and a borrower that covers the borrower's debt obligation in the event of default. The legislative proposals under consideration for a federal credit risk loan guarantee program would lower interest rates and give risk-averse institutional lenders security in their investment. If loan guarantees are combined with ESPCs, where the ESCO takes on the technical and performance risk (possibly backed even further by energy savings insurance), the loan guarantee covers the relatively small risk of owner default. The proposal has been embraced by the CRE industry and spearheaded by the U.S. Green Building Council, the Natural Resources Defense Council and The Real Estate Roundtable.

Regardless of whether PACE financing is available, establishing a federal or local loan guarantee program to cover credit risk can leverage public funding and ramp-up large scale private investment in the CRE sector.

Federal, state or local governments can also leverage significant private investment by establishing (or seeding) loan loss reserve funds. This credit-enhancing mechanism would cover bridge payments to lenders with a default on their hands. In PACE programs with non-acceleration clauses, because only delinquent property tax payments (typically 1-2 years) need to be cured upon default, the bulk of the assessment survives bankruptcy, and the remaining balance and future payments would be assumed by the new property purchaser. Sources of reserve funding are most commonly being developed at the state level. For example, in April 2010, California passed legislation establishing a statewide PACE Reserve Program. This state-financed loss reserve was created with $30 million from the Renewable Resources Trust Fund.

Energy Savings Underperformance Risk Protection

Energy savings insurance (ESI) policies can provide a backstop for energy savings guarantees provided by ESCOs. In exchange for a premium, the insurer agrees to pay over the term of the policy contract any shortfall in energy savings below a pre-agreed baseline, less a deductible. Pricing is usually expressed as a percentage of energy savings over the term of the contract. A percentage in the 3%-5% range, with a 10% deductible, would not be unusual. The premium is paid once, in the first year of operation. However, depending on the project's financing structure, the up-front ESI premium may be rolled into the financing to enable payment over time.(6)

There are a number of benefits associated with ESI. These include:

1. ESI transfers performance risk from the balance sheet of the entity (ESCO) implementing the energy savings project.

2. ESI forces the criteria for defining baseline energy use levels and projecting savings from energy efficiency improvements to be totally transparent and explicit.

3. ESI can result in higher project confidence among building owners desiring to make significant energy efficiency improvements and lenders financing these improvements.

4. ESI can help avoid disputes with ESCOs over energy savings.

5. ESI, as a credit enhancement, can lower the cost of financing.

6. The ESI insurer can provide third-party review of engineering and design specifications and third-party involvement in ongoing energy savings measurement and verification, thereby increasing the building owner's confidence level to invest.

A number of insurance companies are now exploring the ESI concept and market opportunity. One company, Hannover Re, a leading international reinsurance company working with Energi Insurance Services (Peabody, MA) recently launched an ESI product for ESCOs known as the "Energy Savings Warranty." The PACE Commercial Consortium (PCC), created by Carbon War Room, has chosen to incorporate this "Energy Savings Warranty" into their program to reduce the risk. It is expected that other insurers will follow as the market expands and emerging long-term energy retrofit financing programs such as PACE programs take root.

ENERGY EFFICIENCY LOAN UNDERWRITING

No matter what type of financing is ultimately selected to fund an energy efficiency project, it will have to be underwritten. Deep energy retrofits in existing buildings commonly require analysis of the whole building and application of multiple energy conservation measures (ECMs). Underwriting energy efficiency loans for commercial whole building retrofits that involve multiple (and often interacting) ECMs can now be accomplished in a technically sound, consistent, practical and fully transparent manner using an emerging best practice. This technical underwriting process can now provide underwriters with the confidence they need to underwrite energy efficiency investments.

There are three primary industry standards or protocols that are working in combination to accomplish this new high quality underwriting:

1. The ASTM E2797-11 Building Energy Performance Assessment (BEPA) Standard(7) published in February 2011 that focuses on energy data collection and analysis to provide a standardized baseline;

2. The American Society of Heating, Refrigerating and Air-Conditioning Engineers (ASHRAE) Level II and Level III Energy Audit Guidelines(8) to determine the optimized bundle of ECMs and the associated key financial metrics; and

3. The International Performance Measurement and Verification Protocol (IPMVP) guidance document to measure and verify energy savings using the whole-building Option C method.(9)

Baseline Energy Use and ECM Determination

The ASTM BEPA Standard ensures that building energy data collection and analysis provides a firm foundation to establish baseline energy use. Until recently, no consistent standardized methodology existed in energy auditing for the collection and analysis of building energy use data to establish this baseline.

While it may seem relatively straightforward to simply collect utility data, the devil is in the details. For example, prior to the ASTM BEPA Standard, there was no standard time period over which building energy use data had to be collected and energy professionals commonly used anywhere from one to three years. The ASTM BEPA Standard established three years as the time period over which energy use data should be collected, or back to the building's last major renovation if completed in less than three years, with a minimum of one year if reliability criteria can be met. Furthermore, if a building had undergone a major renovation, there was no standard as to how this should be considered, if at all. There was not even a standard definition as to what constituted a major renovation. The ASTM Standard defines a major renovation as one which either involves expansion (or reduction) of a building's gross floor area by 10% or more, or that impacts total building energy use by more than 10%.

Finally, there were no standards on how weather conditions should be analyzed and taken into consideration, how building operating hours should be factored into the analysis, or how building occupancy should be considered. The ASTM BEPA Standard prescriptively addresses each of these issues with the result that use of the ASTM BEPA methodology tightens many of the loose ends in the energy audit guidelines.

The ASHRAE Level II or Level III energy audit then builds on the solid baseline foundation created by the use of ASTM BEPA Standard methodology, identifies energy use by major building component or function, and determines the optimized bundle of recommended ECMs that provide a compelling investment opportunity

Energy Savings Projections

A key to making energy efficiency investment is the ability to project energy savings with a high degree of confidence. To accomplish this, the ASTM BEPA Standard is used in conjunction with the ASHRAE Level II or Level III energy audit results. The projected energy savings after the ECMs are installed is determined by the difference between what the projected energy use would be without the ECMs installed and what the projected energy use would be assuming the ECMs are installed. The former can be determined from the building energy use equations (associated with electricity and fuel) developed using ASTM BEPA methodology using the mean values for the independent variables (historic weather, occupancy, operating hours, etc.) in the equations. The latter is determined by the energy professional conducting the energy audit. For each recommended ECM, the energy savings are projected (including accounting for potential interactive effects) and then deducted from the projected energy use that would have existed assuming the ECMs had not been installed.

To reflect uncertainty in the analysis, it is common to express energy savings in conjunction with confidence and precision levels. Confidence level is the probability that the savings will fall within the precision range (or the range in which the true value is expected to occur). For commercial buildings, error is inherent in the baseline energy use equations (developed from regression analysis) for a number of reasons:

1. Rarely is it possible to identify every independent variable impact-
    ing a building's energy use, particularly those activities associated
    with human activity, i.e., occupant behavior (such as the open win-
    dow in conditioned space or the electric heater in a workspace).

2. Some baseline energy use data rely on delivery invoices rather
    than meters, e.g., fuel oil delivered for heating, which when av-
    eraged over the use time frame may not coincide precisely with
    actual use.

3. Utility invoices may include estimates for a specific period.

4. Electric meters may be misread

     The uncertainty (or error) associated with the building energy use
equations can be determined from the difference between the monthly
actual energy use and the calculated monthly energy use over the 36
months in which energy use data was collected for the baseline. These
errors can then be statistically analyzed and the standard deviation determined. Once a confidence level, and therefore a precision, is specified,
the tolerance around a calculated energy use value can be established. For
example, if a 95% confidence level is specified, this corresponds to 1.96
standard deviations. Hence, the projected energy use range (upper and
lower) at the specified confidence level around each calculated value can
be determined. The end result is a projected (calculated) energy use range
for each month in the desired reporting period, assuming the ECMs had
not been installed. Uncertainty can also be included around the expected
performance of each ECM scheduled to be installed, i.e., related to the
auditor's confidence level around the projected energy savings for each
of the ECMs. Uncertainty is factored into the underwriting analysis to
provide further confidence in the evaluation of projected energy savings.
(4) (Refer to Figure 6-1.)

Measurement and Verification
     The final step in the analysis is measurement and verification (M&V)
of the energy savings with a high degree of confidence after the ECMs
are installed. To accomplish this, the industry relies on the International

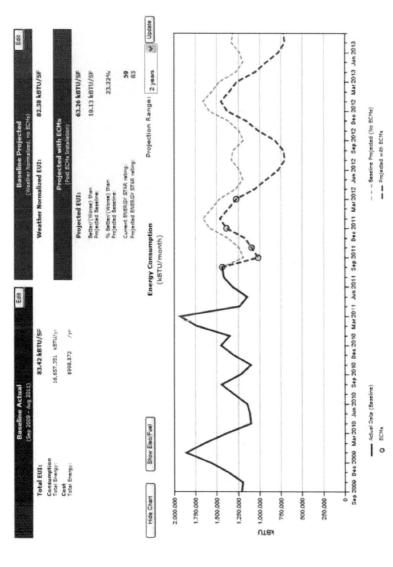

Figure 6-1. Presentation of Projected Energy Savings before Installation of ECMs.

Performance Measurement and Verification Protocol (IPMVP) guidance document. The ASTM BEPA Standard can again be used in conjunction with the IPMVP protocol to facilitate cost effective performance verification after ECMs are installed. The measured and verified energy savings in the desired reporting period (e.g., 12 months after the ECMs have been installed) is determined from the difference between the actual monthly energy use in the reporting period and the projected monthly energy use in this same reporting period assuming the ECMs had not been installed. The former is a measured value. The latter can be determined using the building energy use equations developed in the ASTM BEPA methodology, only this time incorporating the actual monthly values for the independent variables (historic weather, occupancy, operating hours, etc.) in the equation. As before, uncertainty analysis can again be included. ASTM BEPA methodology complements the IPMVP and adds value by providing the necessary depth and prescriptiveness to the pre-ECM and post-ECM evaluation process. (Refer to Figure 6-2.)

EMERGING BEST PRACTICE FOR
ENERGY EFFICIENCY PROJECT FINANCING

In order to implement a successful energy efficiency retrofit project and obtain financing under the most attractive terms, a "best practice" consisting of the following steps is emerging in the CRE market.

Upfront
1.   Conduct an ASHRAE Level II or III energy audit incorporating ASTM BEPA methodology to identify baseline performance and energy savings opportunities.

2.   Identify applicable government/utility grants, rebates and incentives.

3.   Select energy conservation measures (ECMs) meeting criteria (ROI, payback time, etc.).

4.   Determine total project cost and payback time.

5.   Identify projected energy savings at the selected confidence level using ASTM BEPA methodology and the IPMVP framework.

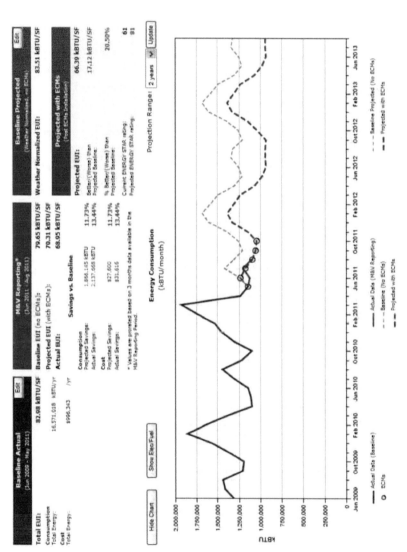

Figure 6-2. Presentation of Actual Energy Savings after Installation of ECMs.

Financing

6.   Establish the amount of financing needed and the preferred payback period.

7.   Obtain the cost of energy savings insurance and a commitment letter from the carrier.

8.   Solicit interest from lending sources by providing a full documentation package, including the ASHRAE Level II or III energy audit report incorporating the ASTM BEPA, and the M&V plan, to support the energy savings projections at the required confidence level.

9.   Secure financing under preferred terms.

Implementation

10.   ECM engineering and design.

11.   ECM installation.

12.   ECM commissioning.

Performance M&V

13.   ECM performance measurement and verification (M&V) relying on the M&V plan and ASTM BEPA methodology within the IPMVP framework.

14.   Conduct annual M&V and provide documentation to the lender, insurer and any other stakeholders.

This emerging best practice for underwriting supports financing for energy efficiency retrofit projects.

CASE STUDIES

The Los Angeles Commercial Building Performance Partnership

The Los Angeles Commercial Building Performance Partnership (LACBPP) program was developed by the City of Los Angeles and the Clinton Climate Initiative and offers capital providers the opportunity to make financially attractive investments in energy efficiency projects in the CRE market. LACBPP has been designed to connect building owners with a range of investors interested in funding an energy efficiency project through a variety of structures, from energy service agreements and equipment leases to innovative PACE financing options.

The innovative PACE financing option is an "open market" model where the city acts as a conduit for private investment (including traditional debt lenders) in energy efficiency retrofit projects. Building owners negotiate financing terms with investors of their own choice. Owner-negotiated terms are then reflected in a loan agreement funded through issuance of a bond by LA County. The bond is sold to the investor that underwrote the deal in a private placement. The existing mortgagee may also underwrite and purchase the bond. Repayment is secured through a contractual assessment in first position on the building's property tax bill (this is subject to the mortgage holder's consent to the PACE assessment). The program offers 100% financing on the total project, at lower interest rates and with longer term financing to allow projects to be cash flow positive from day one. Moreover, since property taxes are an operating expense, the transaction may be considered "off balance sheet." The PACE assessment transfers with the real estate in the event the building is sold in the future.

To prepare a project for investment, the LACBPP model relies on the emerging best practice, including the ASTM BEPA for data collection and analysis, an ASHRAE Level II energy audit to determine ECMs and the IPMVP for energy savings measurement and verification.

The Sustento Group provides program oversight. Willdan is the program manager. Advanced Energy Innovations (AEI) serves as the third-party technical peer reviewer. Sustainable Real Estate Solutions (SRS) provides the technical and financial underwriting software platform used by all parties to ensure underwriting best practice compliance and reporting. Seven energy auditing and engineering firms have been pre-qualified to conduct the technical underwriting scope of work.

PACE Commercial Consortium

The PACE Commercial Consortium uses a "warehouse" PACE model. The consortium was formed by Richard Branson's non-profit Carbon War Room and is the largest single private-sector investment to-date in the commercial property energy retrofit market. The Consortium will provide up to $650 million for retrofit projects in Miami-Dade County, Florida ($550 million) and Sacramento, California ($100 million).

The PACE Commercial Consortium model relies on the emerging best practice, including the ASTM BEPA for data collection and analysis, an ASHRAE Level II energy audit and the IPMVP for energy savings The Consortium is managed by Santa Rosa, California-based Ygrene Energy Fund. Short-term loans will be provided by Barclays Capital. Loans will be warehoused until critical mass is reached at which time they will be bundled into long term bonds (resembling those routinely issued by governmental taxing districts) to be marketed by Barclays. Project management and engineering for the Consortium is handled by Lockheed Martin (the ESCO) and the energy savings insurance will be provided by Energi Insurance Services. Sustainable Real Estate Solutions (SRS) provides the technical and financial underwriting software platform used by the parties to ensure underwriting best practice compliance and reporting.

CONCLUSION

There are a number of financing options available that can provide commercially-attractive funding for energy efficiency retrofit projects in the CRE market. These include PACE tax lien financing, ESCO direct financing or third party financing using the PACE structure, ESA provider financing through private parties or using PACE financing, and bank (private investment) financing through a PACE financing structure. Each of these financing mechanisms has a structure that can enable positive cash flow from day one.

Most of the activity today is in the MUSH and owner-occupied segment of the CRE industry. However, as PACE programs are developed and expand nationally, deeper energy retrofits will be possible in these industry segments, and the much larger multiple-tenant sector of the CRE industry will become a growing target market. Moreover, as the economy improves, the release of pent-up demand to replace outdated, aged energy-consuming equipment in CRE buildings will also contribute to energy efficiency project market demand.

The key to providing financing is an ability to underwrite loans in a standardized, technically sound, consistent and fully transparent manner. A best practice has emerged which relies on an ASHRAE Level II or Level III energy audit and the IPMVP framework for M&V, both supported by ASTM BEPA methodology. The underwriting best practice can provide lenders with confidence in the energy savings projections prior to the installation of the ECMs and confidence that the energy savings can reliably be measured and verified after the ECMs are installed. The best practice has already been incorporated into the Los Angeles Commercial Building Performance Partnership (LACBPP) Program and the PACE Commercial Consortium.

The emergence of this best practice is finally enabling building energy efficiency financing to become a mainstream financial asset class with a high degree of standardization, predictability and scale. It is expected that this best practice will go a long way toward accelerating large-scale adoption of energy efficiency investment in the CRE market.

> *"A best practice has emerged which relies on an ASHRAE Level II or Level III energy audit and the IPMVP framework for M&V, both supported by ASTM BEPA methodology. Underwriting energy efficiency loans for commercial whole building retrofits that involve multiple ECMs can now be accomplished in a technically sound, consistent, practical and fully transparent manner."*

References

1. Buonicore, A.J., "Using the New ASTM BEPA Standard in the Property Transaction Market," Building Energy Performance Assessment News, Critical Issues Series, Paper No. 11-001, August 2, 2011. (www.bepanews.com)
2. Buonicore, A.J., "The Formidable Challenge of Building Energy Performance Benchmarking," Building Energy Performance Assessment News, Critical Issues Series, Paper No. 10-001, April 5, 2010. (www.bepanews.com)
3. Buonicore, A.J., "Energy Efficiency Retrofit Financing Options for the Commercial Real Estate Market," Building Energy Performance Assessment News, Critical Issues Series, Paper No. 12-001, February 15, 2012. (www.bepanews.com)
4. Buonicore, A.J. and Halpin, C.F., "M&V in Energy Performance Contracting Using ASTM BEPA Methodology," Building Energy Performance Assessment News, Critical Issues Series, Paper No. 11-004, December 13, 2011. (www.bepanews.com)
5. New Buildings Institute, "Deep Energy Savings in Existing Buildings: Summit Summary," February 2012 (data extracted from U.S. Energy Information Administration, 2003).
6. Buonicore, A.J., "Energy Savings Insurance and the New ASTM BEPA Standard," Building Energy Performance Assessment News, Critical Issues Series, Paper No. 11-003, November 15, 2011. (www.bepanews.com)
7. ASTM Standard Practice E 2797-11, Building Energy Performance Assessment, published by ASTM, Conshohocken, PA, February 2011.
8. ASHRAE, Procedures for Commercial Building Energy Audits, 2nd Edition, 2011.

Efficiency Valuation Organization, "International Performance Measurement and Verification Protocol, Concepts and Options for Determining Energy and Water Savings," Volume 1, EVO 10000—1:2010, September 2010.

9.   Efficiency Valuation Organization, "International Performance Measurement and Verification Protocol, Concepts and Options for Determining Energy and Water Savings," Volume 1, EVO 10000—1:2010, September 2010.

Acknowledgement

The author would like to acknowledge and thank the following individuals who reviewed this chapter and provided valuable comments and suggestions:

Mark J. Bennett, Esq., Miller Canfield Peter L. Cashman, Managing Director, BP, LLC David Gabrielson, Executive Director, PACENow Kerry E. O'Neill, Sr. Advisor, Clean Energy Finance Center Brian J. McCarter, CEO, Sustainable Real Estate Solutions (SRS).

Chapter 7

How To Make It Easy For The Lender To Work With You
Key Risk and Structuring Provisions for Bankable Transactions

By Jim Thoma
Green Campus Partners, LLC*

INTRODUCTION

Energy efficiency (EE) and distributed generation (DG) projects, in all their varying forms, are technically complex undertakings, requiring significant design, engineering and development efforts to create a transaction that is economically compelling to the end-use customer. However, all these efforts could be for naught if third party financing is required and the transaction is not structured to satisfy the requirements of the debt and equity investors in the capital markets. Few things are more frustrating for the customer, developer, energy services company (ESCO) and other contractors than having high expectations for a project, just to learn that funding is unobtainable due to an improperly structured transaction that cannot clear market with cost effective financing. *This chapter is intended to help project developers, ESCOs, contractors and their customers avoid this problem by identifying and explaining specific risk and structural provisions that are key to creating bankable EE and DG projects.*

Developing a bankable project requires a thorough understanding and balancing of the needs and objectives of the three main parties to the

---

*This material is the work of the individual author and is presented for informational purposes only. This work does not represent the formal opinions or positions of Green Campus Partners, LLC including any of its officers, investors, subsidiaries and affiliates. Readers are strongly encouraged to consult their own independent tax, legal and accounting advisors when entering into any financing or investment transaction.

deal—the customer (end user or obligor), the development team (which may include a developer, ESCO and other contractors) and the investors (which may include both debt and equity). Although each party's objectives and requirements must be considered when developing the deal, this chapter is intended to primarily focus on the debt investor or lender's perspective and, by doing so, explain the critical components of a bankable transaction. Once a project financing is structured to be bankable, the debt capital markets should welcome it with virtually unlimited capacity.

EE and DG financing transactions can take many different forms, as long as they're properly structured, priced and documented. The initial challenge is to accurately identify a structure that meets the needs of all three parties to the transaction. In most cases, the parties begin with a common structure, but work through a series of negotiated modifications to achieve the final bankable result. Many development teams have learned the hard way that leaving the lender out of these structural negotiations usually results in unmet expectations and frustration for the customer.

The matrix in Table 7-1 provides an overview of EE and DG project types, financing solutions and corresponding contract structures commonly used to fund bankable transactions.

Table 7-1

Project Types	Financing Solutions	Contract Structures
Energy Efficiency (Lighting, Controls, HVAC Upgrades, Water Systems, etc.)	**Debt Financing** **Leasing** **Developer Owned (Project Financing)**	• Tax-Exempt Lease • Fed ESPC & UESC • Tax Credit Bonds • Finance or Operating Lease • Energy Savings Agreement (ESA)
Renewable Energy (Solar, Wind, Biomass, Fuel Cell, LFGTE, Small Hydro)	**Debt Financing** **Tax Leasing** **Developer Owned (Project Financing)** (Inverted Lease Pass Through, Partnership Flip, Levered or Unlevered Investment)	• Loan (Note/Security Agreement) • Tax Credit Bonds • Tax-Exempt Lease • Tax Lease • Power Purchase Agreement (PPA)
Distributed Energy or Distributed Generation (CUPs, Combined Heat & Power, District Energy Plants)	**Debt Financing** **Leasing** **Public-Private Partnerships** **Developer Owned (Project Financing)**	• Tax-Exempt Lease • Fed ESPC & UESC • Finance or Operating Lease • Utility Services Agreement (USA)

Figure 7-1 depicts a basic debt financing structure for EE and DG projects.

Figure 7-2 depicts a basic service agreement structure (ESA, PPA or USA) for EE and DG projects.

RISK ASSESSMENT CATEGORIES

Assessing, allocating and mitigating risk is the foundation of a successful transaction development process and serves as the initial point of focus for achieving bankable transactions. In order to attract capital markets investors, a transaction must present a risk profile commensurate with the nature of the financing (i.e. debt or equity) and anticipated investor returns. This concept is frequently and most commonly described in terms of a basic risk/reward analysis—higher risk requires a greater return to the investor, and at some point unmitigated risks will be too daunting to clear

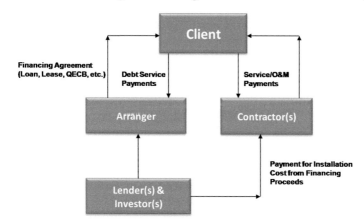

Figure 7-1. Basic Debt Structure

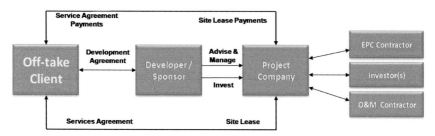

Figure 7-2. Basic Debt Structure

market and obtain funding. This is the foundational structuring concept it its simplest form, but risk analysis and mitigation in energy management transactions is significantly more complex than simply pricing the transaction to provide attractive investor returns.

In underwriting an energy management transaction, the capital markets analyze several broad categories of risk, including:

1.   Obligor Credit Risk
2.   Construction/Installation Risk
3.   Counterparty (ESCO/Contractor/Other) Credit Risk
4.   Technology Risk
5.   Performance Risk
6.   Structural Risk

In order to structure a bankable transaction, each of these risks must be considered and mitigated to the extent required by the (debt) capital markets. To do so, one must understand the nature of these risks and the capital markets' tolerance for each of them.

1. Obligor Credit Risk

Obligor credit risk is the keystone risk for all financing transactions and can be simply stated as the end-user's ability to satisfy all of its obligations (payment and otherwise) under the terms of the transaction. Despite the wide variety of financing structures for EE and DG transactions (as summarized in Table 1 above), their common foundation is the end-user's obligation to make payments in exchange for the products or services received upon funding of the transaction. In whatever form it takes (i.e. lease payment, loan payment, services payment, utility payment, etc.), that payment obligation provides the lender's primary source of debt service, in addition to returns on the equity investment, if any. As a result, the bulk of the lender's underwriting activities are focused on a detailed analysis of the end-user's overall credit profile. Operating performance, profitability, cash flow, debt service capacity, liquidity, balance sheet strength, market/competitive positioning, management capabilities and prospects for future success are among the many factors thoroughly examined during the underwriting process. In order to achieve the desired end result, (i.e. transaction approval and funding), the credit analysis must conclude, with a high probability, that the obligor is capable of fully satisfying its obligations throughout the full term of the transaction.

Credit and risk analytics are also used to determine transaction pricing parameters, such as loss given default, liquidity premiums and risk adjusted return requirements.

2. Construction/Installation Risk

Construction/installation risk is defined as the risk of loss to the investor during the period of time the necessary facilities and equipment comprising the project are being constructed or installed in preparation for commercial operation. Construction/installation risk is a function of the project specifications, the length of the construction period, the nature of the customer, ESCO and other contractor obligations during the construction period and the technical capabilities and financial condition of those parties responsible for constructing and commissioning the project according to the contractually agreed upon parameters.

The assessment of construction/installation risk begins with identifying the party responsible for guaranteeing repayment to investors should the project not be completed on time, on budget or at all. In most cases, construction/installation risk is borne by the ESCO or other prime contractor and or its subcontractors in the form of repayment guarantees to the lender in the event that the project is not completed to the end-user's satisfaction and there is no commencement of debt service payments under the applicable transaction documents. Depending on the credit strength of the party or parties guaranteeing the repayment of funds advanced during the construction period, the lender may also require additional collateral (e.g. cash or letter of credit), step-in rights and/or payment and performance bonds from a creditworthy surety.

3. Counterparty Risk

Counterparty (e.g. ESCO or contractor) risk is a function of the applicable party's role in the project, such as construction and ongoing operations and maintenance of the project. Counterparty risk typically takes three main forms: construction/installation risk, performance risk and general credit risk of insolvency and/or contractual default. Because these counterparty risks can stem from varying yet interrelated circumstances, the lender must analyze them individually and mitigate each to the maximum extent possible.

As previously discussed, the lender must have a reliable source of repayment for all funds disbursed during the construction/installation process. Direct recourse to the ESCO or contractor is the most common vehicle for sufficiently mitigating construction/installation risk. In order to rely on its recourse, the lender will require a thorough review and approval of the ESCO's or contractor's credit profile. These counterparties should be prepared to provide at least three years audited financial statements, construction and project management qualifications, historical performance data and any other information necessary to underwrite the transaction. Depending on the credit strength of the ESCO or contractor, the lender may impose additional credit conditions such as performance bonds and parent guarantees.

4. Technology Risk

In many EE and DG transaction structures, the end-use obligor's (customer's) payment obligations may be partially or wholly contingent on performance of the project. For example, under a typical Utility Services Agreement, the customer is only required to pay for services received (e.g. thermal energy and electricity) in accordance with the terms of the off-take contract. If the equipment comprising the project malfunctions or does not operate as expected, the revenue stream to support debt service will be jeopardized and the lender could suffer a loss. As a result, lenders will perform significant due diligence on the key technology utilized in the project, focusing on issues such as historical performance, uptime reliability, operations and maintenance requirements, market penetration and acceptance, technical standards, mechanical complexity, fuel requirements (including fuel supply and price risk) and other indicators of technical risk. In most cases, the lender will engage an independent engineer to assist in assessing overall project and technology specific risk.

5. Performance Risk

Performance risk is measured in terms of the lender's risk of repayment in the event of a performance deficiency and/or performance default within the financed project. If the end-user has an absolute and unconditional payment obligation to the lender regardless of project performance problems such as equipment failure or service disruptions, then the lender's exposure to performance risk is minimized. On the other hand, if the end-user's payments are conditional upon satisfactory performance as in the case of most traditional project finance transactions, then the lender must have recourse to a third party (typically the ESCO or other operations and maintenance contractor) for debt service shortfalls resulting from project performance deficiencies. In that case, the lender will closely scrutinize performance risk and the performance obligor's ability to satisfy its obligations in the event of actual or alleged non-performance. As mentioned above, additional structural and credit enhancements may be required to fully mitigate performance risk.

6. Structural Risk

Structural risk stems from the contractual nature of the transaction and the specific terms and conditions therein. In order to develop a bankable EE or DG transaction, it is critically important to consider and properly address these risks early in the transaction development process. Failure to clearly identify these risks and allocate them to the appropriate parties is the most common mistake that leads to a non-bankable transaction. The most important of these risks and the most effective ways to mitigate them are identified and discussed in detail below.

Performance Risk vs. Credit Risk

An important general rule of thumb in financing EE and DG projects is that debt investors in the capital markets are in the business of taking credit risk, not performance risk. As a result, transactions must be structured with bright line separation of these risks, with credit risk clearly allocated to the lender and performance risk clearly assumed by the ESCO, contractor or the end-user obligor. This bright line separation enables the lender to evaluate and underwrite the transaction according to standard credit procedures, without focusing on technical specifications, savings estimates, energy costs, equipment reliability and a host of other factors that affect the projects ability to perform as expected and as required under the terms of the applicable contract(s). The bottom line is that the lender must be protected from any risk of non-payment due to an actual or alleged performance deficiency or default. Conversely, if the end-use obligor is simply unable to make debt service payments because of a working capital shortfall or liquidity crunch that is not a result of project non-performance, then that is the lender's risk.

Payment Obligation(s)

Payment obligations are closely related to and sometimes a function of the separation of performance and credit risk. The lender's preferred structure is to have a "hell or high water" payment obligation directly from the end-user or payment obligor regardless of project, ESCO or contractor performance. This is typically the case with basic debt financing and enables the lender to perform a standard analysis of the obligor's creditworthiness and underwrite the transaction if that analysis indicates a high likelihood of repayment. In a structure where the end-use obligor does have the right to suspend payment due to alleged or actual performance deficiencies, the lender will look to the ESCO, contractor or a surety to guarantee repayment. In all cases, the clear separation of performance and credit risk is critical to the lender's ability to preserve the hell or high water nature of the repayment obligation. If done properly, the lender's risk will be limited solely to the creditworthiness of the end-use obligor or, in a performance guarantee scenario, the technical capabilities and creditworthiness of the performance guarantor.

Funding during Construction/Installation

There are a number of ways to provide for funding of the project during the construction or installation period (e.g. escrow funding, construction loans with direct disbursements from the lender, bridge loans, etc.). Regardless of the funding mechanism, the related documents must address specific issues such as mechanics & contractor liens, procedures for invoice submittal, approval and payment, escrow or payment agent costs, title and security interest in the equipment and waivers and consents, if necessary.

Payment & Cash Flow Logistics

Not only do the debt capital markets require a strict repayment obligation, but they also require certainty as to the timing and logistics of those payments. In most cases, the lender or a qualified servicing agent will assume full responsibilities for billing, collection and distribution of payments. When debt service payments are bundled with payments to the ESCO or contractor for its operations and maintenance services (as in Federal Energy Savings Performance Contracts), the lender will require a lock-box agreement for collection and distribution of payments via a pre-defined waterfall agreement. The lock-box would be controlled by the lender under the terms of an agreement that clearly states that all payments received will be applied first to debt service, then to other obligations such as operations & maintenance costs, contractor receivables, cash reserves and lastly to equity, if any.

Taxes, Insurance and Maintenance

If the financing transaction is documented as a standard debt structure or incorporated into an, the lenders will require a "triple net" arrangement. In other words, the lender will require the end user obligor to:

1.   Pay all applicable property, sales and use taxes.
2.   Acquire property and liability insurance in amounts determined by the lender and name the lender as a loss payee and/or additional insured under those policies.
3.   Properly maintain the equipment. The ESCO's operations and maintenance obligations normally cover this requirement.

In a third-party owned project selling output under an executory services agreement (e.g. ESA, PPA or USA) backed by a traditional project financing with equity and debt, taxes, insurance and maintenance will normally be for the account of the project owner or equity sponsor.

Title, Security Interests & Collateral
Depending on the financing structure used, the lender may or may not hold title to the financed equipment. However, in all cases, the lender will require the loan to be collateralized by the financed equipment with a first priority, perfected security interest in the project assets. Three key elements must be included in the documents to satisfy this requirement. They are:

1.   Explicit identification of the assets as personal property and agreement by the parties that the installed equipment will never be deemed fixtures or any other type of real property.
2.   Unequivocal rights of the lender to perfect its first priority security interest in the equipment by making any and all UCC and fixture filings necessary to document its lien(s) on the equipment. Furthermore, the end-use obligor, ESCO and other contractors will be required to keep the equipment free and clear of liens other than those asserted by the lender.
3.   Reasonably limited rights of the lender to access, inspect and/or remove the equipment without liability to the end-user or site owner.

Termination Provisions
Many EE and DG transactions allow for early termination, either at the option of one or more of the parties or upon an event of default. Optional early termination provisions can be accommodated by the capital markets as long as the mechanics of the termination process provide for defined windows of opportunity to terminate, usually with written notice 60 to 180 days in advance, and the required termination payment protects the lender from any loss on its investment. To achieve this acceptable end

result, the transaction must contain unambiguous termination language that provides for payment of an early termination value equal to the lender's outstanding principal balance plus customary early termination fees or a make-whole provision. The fee is usually calculated according to a pre-negotiated sliding scale based on when the early termination takes place—the earlier in the term, the higher the fee. Depending on the preferences of the parties, the early termination value can either be calculated at the time of early termination or be predetermined and documented in an early termination schedule within the financing documents.

In the event of early termination resulting from a performance default, the lender will still require payment of either the defined early termination value or a casualty loss value, as defined in the documents. The structure of the financing will dictate which party is obligated to pay the early termination value. In standard financing transactions directly between the end-use obligor and the lender, the obligor will be required to pay the early termination payment. Depending on the terms and conditions of their contract with an ESCO or other contractor, the obligor may or may not have recourse to the ESCO or contractor if the early termination was caused by a performance default. In bundled service agreement structures, the defaulting party is usually obligated to make the early termination or casualty loss payment, which is assigned to the lender.

Equity/Residual Risk

In some structures, such as operating leases and project financings, the lender or third party investor(s) will make an equity investment in the project. As a result, the obligor's payments will not fully amortize the project cost. The investor must rely on the residual value of the equipment at the end of the term to recoup his original investment plus, ideally, a reasonable return on the invested capital. Due to the limited secondary market for the assets used in EE and DG projects, investors take limited residual risk or simply consider it as potential upside to an otherwise acceptable return if the residual value of the project assets proves to be zero at the end of the transaction term.

Assignment Rights

The lender's unlimited right of assignment is a critical component of a bankable transaction. Without this feature, the deal will be illiquid, isolated from opportunistic trading within the capital markets and extremely difficult, if not impossible, to fund. This reality stems from the fact that all lenders must maintain liquidity in their investment portfolios by having the right to sell (i.e. assign) their paper to other investors in the secondary market. Conversely, liquidity is further enhanced by prohibiting or precisely limiting the assignment rights of the other party or parties to the transaction, especially the end-use obligor. This is a necessary recognition of the fact that the lender's investment decision is based on the creditworthiness of the original obligor. By controlling the obligor's right of assignment, the lender can prevent credit, portfolio and risk-adjusted return deterioration which would result from assignment to and assumption of the borrower's obligations by an entity with a weaker credit profile than the original obligor.

Contracts & Documentation

Once the parties have agreed upon the basic business and structural terms of the transaction, the deal must be documented in compliance with the standards and expectations of the capital markets, most of which have been identified above. An experienced legal team is key to achieving this requirement. Among the most common mistakes that must be avoided are:

Lack of Clarity—the parties' rights and obligations must be stated clearly and unequivocally so as not to be open to interpretation, either by the parties themselves or in a court of law. Something as intuitively simple as the customer's payment obligation can create significant funding problems if not written with the clarity necessary to ensure consistent interpretation by all parties.

Co-mingling of Lender and ESCO or Contractor Risks—the documents must achieve a bright line separation between the lender's credit risk and other project risks such as construction or ongoing performance risk.

Inseparability of Financing and Services—in a services agreement supported by traditional project financing (i.e. where the lender does not have recourse to the end-user or off-taker from the project), the terms and conditions of the financing must be separable from those of the technical project in order to successfully monetize the transaction.

Failure to Include Key Capital Markets Provisions—documents without the previously discussed critical provisions, such as lender assignment rights, security interests in the equipment, hell or high water payment obligations, etc. will be poorly received by the capital markets and difficult, if not impossible to fund in the tight credit environment of today's capital markets.

CONCLUSION

The capital markets have a theoretically unlimited capacity to fund creditworthy, properly structured energy efficiency and distributed generation transactions. By being aware of investor requirements and working collaboratively with a knowledgeable and experienced financing partner from the earliest stages of a project, end-user obligors and their contractors can avoid costly project delays, maintain competitive advantages, meet or exceed customer expectations, avoid leveraging their balance sheets and gain broad access to the capital markets to fund their bankable projects.

Part III

Extra Information that Can Make the Difference between "Yes" and "No"

Chapter 8

How Energy Projects
Improve Stock Prices

John R. Wingender,
Eric A. Woodroof, Ph.D.

Editor's Note: **Energy managers need all the help they can get to implement energy projects. Authors Wingender and Woodroof describe a new way to catch top management's attention, via an objective close to their hearts: How to improve the price of the company's stock.**

The potential for increased profits via *cost-reducing* energy management projects (EMPs) exists in nearly all firms. However, when allocating capital, priority is often given to *revenue-enhancing* projects, such as starting new product lines or joint ventures.

Frequently, these projects are perceived to be superior to EMPs, even though they may yield the same increased profit and present value. A justification is that *revenue-enhancing* projects are more likely to attract publicity and investor attention. Investor speculation and reaction to announcements can increase the firm's stock price. Most EMPs do not generate as much publicity as joint ventures or new product lines.

If "publicity-gaining" potential is a decision factor during project selection, then a new product line or joint-venture will usually be selected over an EMP. But is this a fair comparison? There has not been any research to determine if an EMP announcement increases a firm's stock price. In theory, it should—because most EMPs increase profits (via *cost reduction* instead of *increased revenues*). From a cash flow perspective, an EMP is equivalent to any other profit-enhancing project.

This article seeks to determine whether an EMP announcement correlates with an abnormal increase in a firm's stock price. If such announcements positively impact stock price, then the firm has one more incentive to implement EMPs.

LITERATURE REVIEW

The purpose of this literature review is three-fold:

1.   To demonstrate that EMPs are credible investments, with relatively low risk;

2.   To present some background on stock price reaction to announcements of typical capital investments; and

3.   To show that abnormal increases in stock prices from EMP announcements have not been measured.

Public announcements (such as mergers, joint ventures or new product lines) correlate with abnormal stock price returns.[1,2] When a firm announces a joint venture (or other revenue-enhancing project) it is trying to attract publicity, which can raise the stock price based on expected future profits. However, since such projects can also be unprofitable, the anticipated cash flows are at risk.

When firms implement EMPs, they also expect improved profits by becoming more cost-competitive. EMPs and equipment replacement projects usually have more predictable cash flows (less risk) than many other types of capital investments, especially new product lines or joint ventures.[3] Today, the risk from most EMPs is so low there are many third party lenders who are eager to locate and finance EMPs.[4] In 1995, leasing (including third-party leasing and performance contacting) accounted for nearly one third of all equipment utilization.[5]

Thus, EMPs and other facilities improvement projects are recognized as credible investments; however, they are frequently put on the "back burner" relative to *revenue-enhancing* projects.

Maximizing stock price should be a goal of the corporation. Increasing productivity, offering new product lines, and increasing profits are examples of tangible factors that can increase the firm's stock price. However, stock price may also increase due to intangible factors, such as investor speculation and reaction immediately following an announcement. Executives may incorporate this investor reaction when deciding which projects to implement.

Although investor reaction has not been assessed for EMP announcements, there has been some research in this area. It has been shown that firms increasing expenditures on general facility and equipment improvements had a 1.98% abnormal stock increase immediately after the announcement.[6]

Announcements of joint ventures correlated with a positive abnormal return of 1.95% over a 21-day interval (–10 days to +10 days) centered on the announcement day (day 0).[7] However, not all joint ventures correlate with positive abnormal stock returns.[8]

The correlation between EMP announcements and stock price has not previously been investigated. This article examines whether EMP announcements correlate with positive abnormal stock price returns. If they do, then perhaps the capital budgeting process should incorporate this benefit.

METHODS

Using the Nexis/Lexis Data Base, the world wide web, and other resources, a search for EMP announcements resulted in over 5,500 citations. Of the 5,500 citations, only 23 announcements fit the following criteria:

1. The firm announcing the EMP was publicly traded and its returns were available on the data files of the Center for Research in Security Prices (CRSP).

2. The announcement was the first public information released about the EMP.

3. The EMP was large enough to represent a significant investment for the company. *For example, if a large fast-food chain was announcing an EMP at only one restaurant, it was deleted from the sample. However, if the EMP was company-wide, it was included in the sample.*

4. The announcement was made in a major US newspaper, newswire service or a monthly trade magazine between 1986 and 1995.

The 23 announcements which represented the "Complete Sample," consisted of several subsamples:

- 16 Announcements made within daily newspapers or via electronic wire ("daily")

- 3 Announcements within monthly trade magazines ("monthly")

- 4 Announcements of post-implementation results ("post-imple-mentation")

An additional sub-sample was created ("daily + monthly") to maximize the sample size of preimplementation announcements.

For each announcement, the firm's name, CRSP identification number and announcement date were entered into the *Eventus* computer program, which tracks daily stock price performance for every U.S. stock listed within the CRSP.[9] *Eventus* uses a common "event-study" methodology to calculate abnormal stock returns, and it indicates statistical significance levels. Statistically, a null hypothesis (H_0) was proposed and tested against an alternative hypothesis (H_a).

H_0 = EMP announcements have no impact on stock price.

H_a = EMP announcements have a positive impact on stock price.

Analysis Intervals
Usually, a firm's stock price performance is analyzed over several time intervals around the announcement date. Often, a stock price improvement can be noticed between the day of the announcement (day 0) and the next trading day (day 1). In this interval, the range is represented by the following notation: (day 0, day I) or (0,1). Another typical interval for event-study analysis is a two-day interval, one day before the announcement to the announcement date (–1,0).[10]

EMP announcements frequently appear in monthly trade magazines, which are distributed to readers on different days in different geographic locations. In this case, an exact announcement date cannot be determined. Thus, analysis of the stock performance over a wider interval is appropriate. The (–10,10) interval represents the period at which EMP announcements would most likely be noticed through monthly trade magazines.

In addition, since EMP announcements may not capture as much publicity as other announcements, it is reasonable to expect a longer period for the market to "learn" about an EMP.

The (–10,10) interval is useful for identifying if an abnormal stock price increase correlates with an EMP announcement. However, to observe the long-term stock impact, the sub-samples were analyzed over additional intervals, such as (1,100), (1,150), etc.

Applying the aforementioned hypothesis tests to the sub-samples yielded the results which are presented in the following section of this article. A more detailed explanation of the "event-study" methodology (statistical analysis) is included in the Appendix.

RESULTS

Tables 8-1 and 8-2 present the short-term and long-term abnormal returns. The returns are categorized by interval around the announcement date. The level of significance at which H_0 was rejected is also indicated.

The daily + monthly sub-sample is the most appropriate sample because it is the largest sample possible that excludes post-implementation announcements. The post-implementation subsample is substantially different in nature because it represents firms announcing cost savings (increased profits) from projects already implemented.

Using the daily + monthly sub-sample, EMPs correlate with a 3.90% increase in stock price, measured from ten days prior to the announcement to ten days after the announcement, (–10,10). The level of significance was 0.01. See Table 8-1.

Table 8-2 shows the long-term performance, where the daily + monthly sub-sample correlated with a 21.33% abnormal return over the (1,150) interval, at the 0.001 significance level. Figure 8-1 is a graphical illustration of the abnormal returns over the long-term interval.

CONCLUSION

The results from this study indicate that EMP announcements do correlate with significant abnormal increases in a firm's stock price. On average, an EMP announcement correlated with a 21.33% abnormal increase in the firm's stock price. This increase was experienced from the day after the announcement to 150 days after the announcement. This increase is *in addition to* the risk-adjusted return the firms would normally experience.

For example, during a "bull market" a firm 's expected return was 10%.

Table 8-1. Abnormal Return of Firms Announcing Energy Management Projects

Samples	# of Firms in Sample	Day Range From Announcement Date			
		(-5,5)	(-10,10)	(-15,15)	(-20,20)
Complete Sample	23	2.21%, *	2.80%, **	1.06%, *	1.92%, *
Announcements from Daily Newspapers	16	2.89%, **	3.75%, ***	2.31%, **	3.18%, **
Announcements from Monthly Magazines	3	2.31%	6.91%	8.23%	8.50%
Post Implementation Announcements	4	-0.57%	-2.70%	-5.86%	-7.12%
Daily + Monthly Subsamples Combined	19	2.43%, *	3.90%, ***	2.83%, *	3.64%, **

Significant at: * =0.10
             ** = 0.05
             *** = 0.01

Table 8-2. Long-Term Abnormal Return of Firms Announcing Energy Management Projects

Samples	# of Firms in Sample	Day Range From Announcement Date					
		(1,100)	(1,120)	(1,150)	(1,200)	(1,220)	(1,240)
Complete Sample	23	8.23%, *	11.00%, **	13.93%, ***	12.09%, **	13.75%, **	12.71%, **
Announcements from Daily Newspapers	16	10.67%, **	16.29%, ***	17.66%, ***	13.38%, **	17.10%, **	17.11%, **
Announcements from Monthly Magazines	3	27.05%	21.38%	31.54%	44.89%	42.44%	53.02%
Daily + Monthly Subsamples Combined	19	14.07%, **	18.03%, ***	21.33%, ***	21.73%, ***	23.42%, ***	24.95%, ***

Significant at: * = 0.10
             ** = 0.05
             *** = 0.01
             ****= 0.001

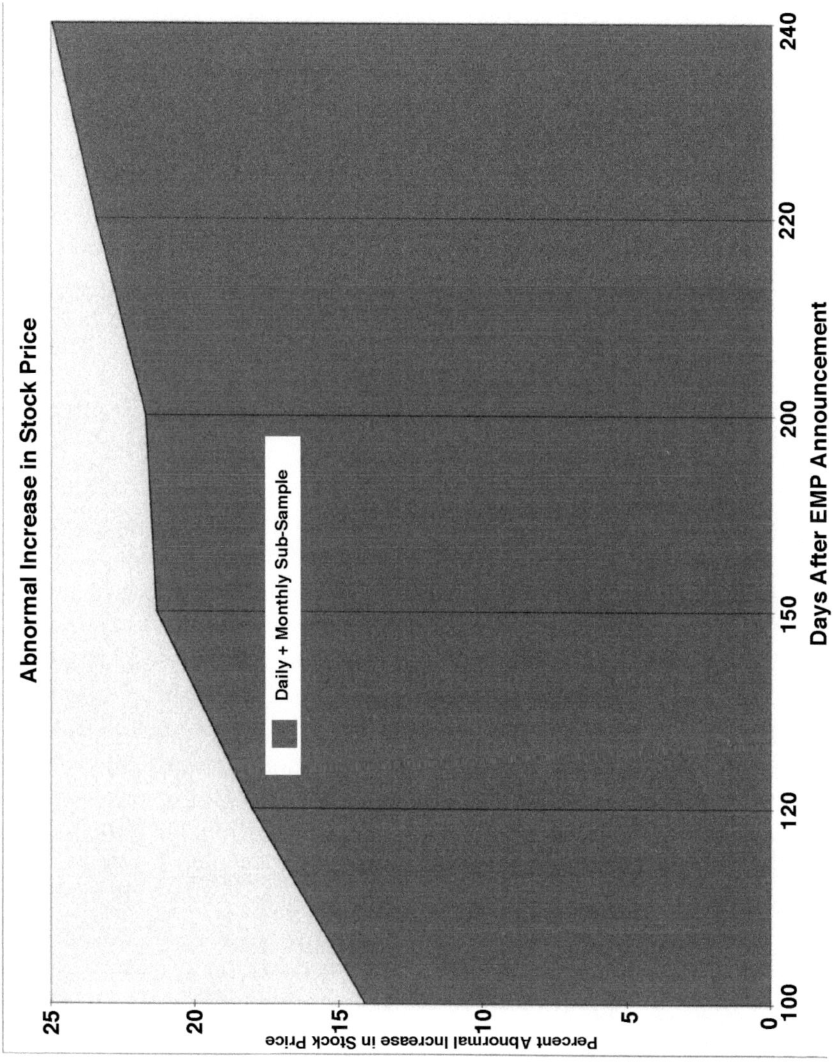

Figure 8-1. Abnormal Return of Firms Announcing Energy Management Projects

After the announcement, the return increased by 21.33%, for a net return of 31.33%. Because these EMPs were announced by a diverse group of firms at various periods over a ten-year time span, the significance of these results is impressive. In other words, the EMP is probably the only event that all firms within the sample have in common.

From these results, it appears that shareholders recognize EMPs as investments that should increase profits and add value to the firm. With the new information presented here, firms may have an additional strategic incentive to implement EMPs.

DISCUSSION AND RECOMMENDATIONS
FOR FURTHER RESEARCH

Despite the small scale of this study, the significance of the results is impressive. This study could serve as a first step to understanding investor reaction to EMP announcements. Additional studies with increased sample size and greater stratification would yield more information.

It is interesting to note that detailed cost savings estimates were not always included in the EMP announcements. *For example, many firms simply announced that they were going to retrofit a portion of their facilities, without an estimate of dollar savings.* Perhaps more detailed information was released after the announcement date, triggering greater stock price increases in the long-term intervals. However, it is more likely that shareholders associate EMPs as effective profit enhancing projects that are almost always good for the bottom line.

It would be interesting to determine if there is a relationship between an EMP's potential profits and the value of the abnormal return. The value of the abnormal return should predictably be related to the amount of increased profit from the EMP. Identifying these values could indicate whether the investor reaction is proportional to the potential added value of the EMP. Calculating these values would require additional information about each firm as well as each project. This could be a focus of additional research.

It was recognized that the type of EMP could influence the magnitude of the abnormal stock increase. Thus, the Complete Sample was further stratified into two sub-samples: EMPs that were lighting retrofits and EMPs that were installations of other types of energy efficient equipment (such as HVAC upgrades, chiller upgrades, etc.).

Both sub-samples were analyzed by the *Eventus* software. The energy efficient equipment sub-sample correlated with a 1.42% increase over the (–10,10) interval, at the 0.01 significance level; the lighting retrofit sub-sample yielded no significant returns over the (–10,10) interval.

However, because this sub-sample only contained seven firms, this comparison needs to be re-evaluated with a greater sample size. In addition, this analysis was tainted because the post-implementation announcements were included.

It was also recognized that the finance method for each EMP could influence the magnitude of the abnormal stock increase. Since off-balance sheet financing (leasing) is common for EMPs, a comparison was made between EMPs that utilized leasing versus EMPs where the equipment was purchased by the firm (and the debt was carried on their balance sheet). Assuming that stock analysts frequently look at balance sheets to assess a company's performance, it is reasonable to hypothesize that off-balance sheet financed EMPs would correlate with higher abnormal stock returns than EMPs where equipment was purchased by the firm.

The complete sample was further stratified and analyzed by the *Eventus* software. The EMPs that were purchased directly by the firm did show a 3.74% increase over the (–10,10) interval, at the 0.01 significance level. The sub-sample of leased EMPs yielded no significant returns over the (–10,10) interval, although it should be noted this sub-sample included only three firms.

Again, the samples were tainted with the post-implementation announcements. Therefore, this comparison needs to be re-evaluated with a larger sample size, and with the post-implementation announcements removed.

Although the post-implementation sub-sample was small, it did not yield any significant positive abnormal returns. In fact, the returns were negative, although not significant. This is intriguing because a post-implementation announcement is basically a statement of increased profits already realized.

A more extensive study on the effects of post-implementation announcements could reveal if these types of announcements yield different abnormal returns than announcements prior to implementation.

All of the sub-samples should be analyzed over a longer time interval. In addition, increasing the sample size would also improve the validity of the results.

References

1. McConnell, J. and Nantell, T. (1985), "Corporate Combinations and Common Stock Returns: The Case of Joint Ventures," *The Journal of Finance*, **40** (2), 519-536.
2. Pettway, R. and Yamada, T. (1986), "Mergers in Japan and their Impacts on Stockholders' Wealth," *Financial Management*, **15** (4), 43-52 (Winter 1986).
3. Pohlman, R., Santiago, E., and Markel, E. (1988), "Cash Flow Estimation Practices of Large Firms," *Financial Management*, **17** (2), 71-79.
4. Zobler, N. (1995), "Lenders Stand Ready to Fund Energy Projects," *Energy User News*, **20** (3), 19.
5. Sharpe, S. and Nguyen, H. (1995) "Capital Market Imperfections and Incentive to Lease," *Journal of Financial Economics*, **39** (2), 271-294.
6. McConnell, J. and Muscarella, C. (1985), "Corporate Capital Expenditure Decisions and the Market Value of the Firm," *Journal of Financial Economics*, **40** (3), 399-422. *Abnormal return was measured over a two-day announcement period that encompassed the day on which the announcement appeared in print, plus the previous day (–1,0). The stock performance over a 21-day interval (–10,10) was not reported. This abnormal return value can be different than the value over the two-day interval. The result was different from zero at the 0.01 statistical significance.*
7. McConnell, J. and Nantell, T. (1985), "Corporate Combinations and Common Stock Returns: The Case of Joint Ventures," *The Journal of Finance* **40** (2), 519-536.
8. Lee, I. And Wyatt, S. (1990) "The Effects of International Joint Ventures on Shareholder Wealth," *The Financial Review*, **25** (4), 641-649.
9. Cowan, A. (1996), *Eventus Version 6.2 Users Guide*, Cowan Research.
10. Brown, S. and Warner, J. (1985) "Using Daily Stock Returns," *Journal of Financial Economics*, **14** (1), 3-31.

APPENDIX

Event-Study Methodology

An "event study" is a popular analysis tool for analyzing stock price reactions to particular events. In this study, the "event" is the announcement of an EMP by one of the sample firms. The event date is the first trading day that the market could react to the announcement. The impact of EMP announcements on stock price is tested by calculating risk-adjusted abnormal returns on and around the announcement date.

For this study, we use the market model event-study method and test the results for significance with the standardized residual method. The market model event-study method uses a linear regression to predict stock returns; then it compares the predicted value to its actual returns.

The abnormal return (ABR_{jt}) is the difference between the actual return (R_{jt}) on a specific date and the expected return ($E(R_{jt})$) calculated for the firm on that specific date. The expected return is calculated using the parameters of a single index regression model during a pre-event estimation period. The regression model parameters are determined by the following equation:

$$R_{jt} = a_j + b_j R_{mt} + e_{jt}$$

where

R_{jt} = the return on security j for period t,

a_j = the intercept term,

b_j = the covariance of the returns on the jth security with those of the market portfolio's returns,

R_{mt} = the return on the CRSP equally-weighted market portfolio for period t, and

e_{jt} = the residual error term on security j for period t.

To calculate the market model parameters (a_j and b_j) a 220-day estimation period was used that begins 260 days before the announcement date. For each sample firm, the event period begins 30 days before the announcement date and ends 30 days after the announcement date. The expected return ($E(R_{jt})$) is then calculated using the return on the market (R_{mt}) for the specific event period date:

$$(E(R_{jt}) = a_j + b_j R_{mt}$$

The abnormal return *(ABRjt)* for an event date is then calculated by subtracting the expected return (which uses the parameters of the firm from the estimation period and the actual market return for a particular date in the event period) from the actual return (R_{jt}) on that date. The equation is as follows:

ABRjt = Rjt - $E(R_{jt})$

The average abnormal return (AAR_t) for a specific event date is the mean of all the individual firms' abnormal returns for that date:

$$AAR_t = \sum_{j=1}^{N} \left(ABR_{jt} \right) / N$$

where N is the number of firms used in the calculation.

The cumulative average abnormal return (*CAAR*) for each interval is calculated as follows

$$CAAR_{T_1T_2} = \sum_{T_1}^{T_2} (AAR_t)$$

The standardized residual method is used to determine whether the abnormal return is significantly different from zero. The standardized abnormal return (SAR_{jt}) is calculated as follows:

$$SAR_{jt} = ABK_{jt}/s_{jt}$$

where

s_{jt} = the standard deviation of security j's estimation period variance of its ABR_{jt}'s.

The estimation period variance s^2_{jt}, is calculated as follows

$$s^2_{jt} = s^2_j \left[1 + 1/D_j + \left[(R_{mt} - \bar{R}_m)^2 / \sum_{k=1}^{D_j} (R_{mk} - \bar{R}_m)^2 \right] \right]$$

$$s^2_j = \left[\sum_{k=1}^{D_j} (ABR^2_{jk}) \right] / (D_j - 2)$$

Where

$\bar{R}_m$ = the mean market return over the estimation period, and

D_j = the number of trading day returns (220) used to estimate the parameters of firm j.

Finally, the test statistic for the null hypothesis (H_0) that $CAAR_{T_1, T_2}$ equals zero is defined as follows:

$$Z_{T_1,T_2} = (1/\sqrt{N}) \sum_{j=1}^{N} \left(Z^j_{T_1,T_2} \right)$$

where

$$Z_{T_1,T_2} = \left(1/\sqrt{Q^j_{T_1,T_2}} \right) \sum_{t=T_1}^{T_2} (SAR_{jt})$$

and

$$Q^j_{T_1,T_2} = (T_2 - T_1 + 1) \frac{D_j - 2}{D_j - 4}$$

Chapter 9

Energy Conservation Also Yields:
Capital, Operations, Recognition and Environmental Benefits

"CORE" Benefits are Highly Probable and Worth a Double-digit Improvement to Energy Savings

Eric A. Woodroof, Ph.D., CEM
Wayne C. Turner, Ph.D., PE, CEM
Warren Heffington, Ph.D., PE, CEM
Barney Capehart, Ph.D., CEM

ABSTRACT

Previous research indicates there are additional (often unreported) benefits from saving energy.[1,2] This chapter identifies these "additional benefits" and describes how to calculate their value.[3,4] In addition, we found that a high percentage of facility managers experienced some of these benefits. For example, in a recent survey 92% of facility managers experienced reduced maintenance material costs as a result of energy conservation (primarily because lights, filters, and other equipment lasted longer when operated less hours per year). Due to site-specific factors, not all facility managers will experience every benefit, however a high percentage of respondents (92%, 71%, and 63%) did experience three of the six additional benefits in the survey. Because facility managers do receive some of these additional benefits, we developed two approaches to quantify their value. When applicable, these benefits should yield a direct and verifiable dollar savings a majority of the time. Via a simple example, we calculated these benefits to be worth approximately 31% of additional value beyond the direct energy dollar savings (and that was only applying half of the possible benefits). There are other benefits that defy quantification, some of which we list at the end of the chapter for use in future research and when evaluating energy conservation projects and programs.

INTRODUCTION

During a diet, when a person does not "over eat," they are likely to receive benefits that extend beyond weight loss. The person may be happier, live longer, have increased stamina and avoid other disease as well as medications. Similarly, the benefits of putting a building on an "energy diet" extend beyond the reduction of energy expenses. Such benefits can include very tangible and measurable values such as reduced material and labor costs, as well as other "soft" benefits such as increased productivity and morale. (*Such soft benefits are not calculated within this chapter.*)[5]

This chapter quantifies the additional tangible benefits from demand side energy conservation activities (reduced operational hours).[6] If a benefit was difficult to quantify, we still identified it, but did not estimate savings. *Note to readers: if you see additional ways to quantify benefits, or if there are other benefits we did not include, please contact the primary author (Dr. Woodroof), who will continue this research.*

EXAMPLES OF ADDITIONAL BENEFITS
FROM ENERGY CONSERVATION

The "additional benefits" are explained with examples to illustrate the value that they represent. The benefits we identify below usually apply to a facility's capital, operations, maintenance, administrative, marketing, environmental and/or other budgets.

1.   Reduced Maintenance Material Costs[7]
Example: If lights are not on as many hours, they may not burn out as often… meaning you will not have to buy as many replacement lamps in a year. Another example is reduced heating, ventilating and air conditioning (HVAC) filter replacement costs as a result of operating such systems fewer hours per year.[8]

2.   Reduced Maintenance Labor Costs[9]
Example: If an HVAC system is not on as many hours, the filters can be

changed less often, resulting in labor savings. Similarly, if lighting is used fewer hours, then lamp lives are longer and annual relamping labor costs decline. Finally, if an energy conservation program results in a labor savings worth 10% of a maintenance person's time, that is a real savings, as the 10% is available for other activities.[10]

3.   Avoided Capital Investment

Example: You reduce energy consumption so much that you don't need to purchase equipment that you thought you needed (an additional chiller, boiler, lights, etc.) This would be a one-time savings, and the amount can be huge. (New chillers are expensive, for example.)[11]

4.   Avoided Procurement Costs

Example: You operate the equipment less hours per year and it lasts longer. Then you don't need to replace it as often, which represents deferred planning, legal, administrative, procurement, and other costs. The savings would depend on the amount deferred and other factors, such as your interest rate, opportunity cost, staff time or new positions avoided, etc.

5.   Avoided Purchases of Carbon Offsets[12]

Example: If your facility is buying carbon offsets, then irt will need to purchase fewer carbon offsets as a result of energy savings. (This is only a direct, measurable savings if your organization is purchasing offsets as an objective to be more sustainable.)[13]

6.   Enhanced Image, Public Relations or Recognition[14]

Example: If your facility pays for public relations or "green" marketing, your success at energy conservation may provide some benefits such as free press in the newspaper, awards, etc.

7.   Reduced Sales Taxes/Environmental Penalties

Example: A company will not pay taxes or environmental surcharges on energy that it does not use. This avoided cost represents an additional 8-15% of the energy savings usually estimated by engineers.[15,16] (Note that if a facility calculates savings using an "average cost per kWh," when the total electric bill is divided by total kWh, these savings are automatically recognized, and the value from Benefit #7 would be zero.)

8.   Improved Building Value

Example: From a property management standpoint, when comparing two identical buildings, if one has reduced operational costs, then that building is worth more. You can estimate the increased value by applying a capitalization rate factor, which equals yearly income/total value. Consider a business that earns $100,000 per year and is valued at $1,000,000—the capitalization rate is 10%. Thus, if that business reduces its operating cost (via energy conservation) by $20,000/year (thereby improving income by $20,000/year), then the total value becomes:

yearly income/cap factor = $120,000/0.1 = $1,200,000

Within this chapter's context, the increased value of a building equals energy savings($)/capitalization rate.[17] Thus, if a building's capitalization rate is 10%, it is reasonable to say that its value is increased by 10 times the operating cost savings (from energy savings).[18]

Although not included in this chapter, it is noted that stock prices of corporations have been proven to improve dramatically when energy management programs are announced, or when an organization publishes its corporate sustainability report.[19,20] It is also worth mentioning that avoided energy expenses go directly to the "bottom line" and result in a very efficient use of money, sometimes even greater than the host company's profit margin.[21]

METHODS TO FIND THE PROBABILITY THAT SPECIFIC ADDITIONAL BENEFITS ARE RECEIVED BY FACILITIES[22]

A survey of 182 energy managers from 182 different organizations was conducted during a 10-day period from late February to early March 2012. To be eligible to participate and receive the survey questionnaire, all participants had to have achieved at least a 20% savings from their energy consumption baseline. Of the 182 energy managers that were sent the survey, 63 facility managers were able to respond to the web based questionnaire. The 63 respondents were primarily from educational facilities across the US. These 63 facility managers provided the results relating to probabilities that are presented in Table 9-1.

Because this survey's participants operated nonprofit or government buildings, our questionnaire did not inquire about *Benefits #7 and #8* because they would not apply to these types of facilities. However, *such benefits can be substantial and would be highly probable in private companies,* which are more likely to sell buildings as well as pay taxes.

For all organizations (public and private), even more additional benefits from energy conservation practices are listed near the end of this chapter.[23]

SURVEY RESULTS

As Table 9-1 illustrates, it is probable that facilities will experience some additional benefits from energy conservation.[24]

PROCEDURES TO CALCULATE ADDITIONAL BENEFITS

Quantifying the value of additional benefits can be accomplished a number of ways, depending on site-specific factors. If it is possible to collect actual savings data, that is the best choice. However, as these benefits occur outside of utility budgets, such data might not be currently tracked within your facility. Therefore, below are two independent calculation approaches, which can be used according what information is available within your organization.

- APPROACH #1: Calculating Benefits Related to a Specific Energy Conservation Measure
- APPROACH #2: Calculating Benefits using Organization-wide Budget Data

Table 9-1. Percentage of Facility Managers that Experienced each Additional Benefit

Additional Benefits of Energy Conservation	% of Facility Managers that Experienced this Benefit
1. Reduced Maintenance Material Costs	92%
2. Reduced Maintenance Labor Costs	71%
3. Permanently Avoided Capital Investment	33%
4. Avoided Procurement Costs	63%
5. Avoided Purchases of Carbon Offsets	10%
6. Enhanced Image, Public Relations or Recognition	44%
7. Reduced Sales Taxes/Environmental Penalties	Not Surveyed
8. Improved Building Value	Not Surveyed

Although both approaches can be applied, in the absence of site-specific budget data, Approach #1 may be able to accurately predict savings using a logical approach and standard industry cost estimates. Approach #2 is more likely to be useful when the facility manager has access to (or can validate) site-specific budget data. Approach #2 also involves far fewer calculations than Approach #1, but it is recommended only when a facility manager has access to budget data.

Within each of the approaches, sample calculations are provided for the additional benefits. *Although not all of the benefits will apply to every facility, the calculations below will guide the reader to estimate the value of each benefit and determine if it applies.*

APPROACH #1: How to Calculate Benefits
Related to a Specific Energy Conservation Measure

This approach involves calculating each benefit associated with an energy conservation measure (ECM) and then adding the benefit values together to determine the *total additional value* per ECM. For example, if you optimized a building system and it lasted longer, you could use the sample calculations below to estimate the values of avoided material, labor, etc. Then you would add these values together to determine the ECM's total additional value. If you had implemented multiple ECMs, you would repeat this approach for each ECM and then use the sum to estimate the total additional value to the facility. Note that all cost estimates can be adjusted to reflect your local conditions.

To illustrate the total additional value from an ECM, we will apply this approach to a simple example and show calculations for each individual benefit.

The Sample Energy Conservation Measure

Consider a lighting system that has 10,000 fluorescent lighting fixtures, each with two lamps and one ballast. Each fixture consumes 60 watts. The baseline operational hours are 5,000 per year and energy costs are $0.10/kWh. Thus, our baseline energy consumption is:

(10,000 fixtures)(5,000 hrs/year)(0.06 kW/fixture)
= 3,000,000 kWh/year

Thus, at $0.10/kWh the annual energy cost to operate the lighting system = $300,000/year.

If we implement an ECM that turns the lights off 25% of the time, then we would save 750,000 kWh/year, or $75,000 per year in direct energy savings. Using the calculations below, we can calculate the additional benefits that extend beyond the energy savings. Although not all additional benefits will apply to this particular ECM, we will show calculations as examples. At the end of Approach #1, we will tally the value of the applicable additional benefits for this ECM.

Benefit #1: Sample Calculation for
Reduced Maintenance Material Costs:

Assume you turn off a lighting system 25% of the time (due to vacancy). If lights are used 25% less, the lighting ballasts (and lamps) should last about 25% longer. Let's calculate the impact on the ballast material first:

A ballast life is rated for 60,000 hours of operation. If your building operates the lights 5,000 hours per year, ballasts would need to be replaced at the 12th year. If there are 5,000 ballasts, each costing about $20 in material (includes shipping and taxes), then at the 12th year the material replacement cost would be:

($20/ballast)(5,000 ballasts) = $100,000

The annualized ballast material replacement cost would be:

($100,000)(1/12 years) = $8,333/year

If the lights are only on 3,750 hours/year (a 25% reduction), the ballasts should last 16 years. This would reduce the annualized ballast material replacement cost to:

($100,000)(1/16 years) = $6,250/year

Thus, the annualized material savings for ballasts are:
$8,333/year – $6,250/year = $2,083/year

Now, we can use similar calculations to quantify the reduced maintenance material costs for the lamps:

Assume there are 10,000 lamps, each costing $2.50 (includes shipping and taxes) and lasting 20,000 hours. If lamps are on 5,000 hours per year, then after 4 years they would need to be replaced, and this would cost:

($2.50/lamp)(10,000 lamps) = $25,000

The annualized lamp material replacement cost would be:
($25,000)(1/4 years) = $6,250/year

Again, if the lights are only on 3,750 hours/year (a 25% reduction), the lamps should last 5.3 years.[25] This would reduce the annualized lamp material replacement cost to:

($25,000)(1/5.3 years) = $4,717/year

Thus, the annualized material savings for lamps are:

$6,250/year − $4,717/year = $1,533/year

Therefore the total annual avoided maintenance material costs (lamps and ballasts) are:

$2,083/year + $1,533/year = $3,616/year

This same approach could be used to calculate maintenance material savings values for other ECMs that extend the lives of motors, filters, etc.

Benefit #2: Sample Calculation for
Reduced Maintenance Labor Costs:
Continuing with the lighting ECM from above, if the lights are used 25% less, the ballasts and lamps should last longer and won't need to be replaced as often, resulting in a labor savings. Let's calculate the impact on the ballasts first.

A ballast life is 60,000 hours of operation. If your building operates the lights 5,000 hours per year, ballasts would need to be replaced at the 12th year. Assume it requires maintenance about 30 minutes to replace a ballast, including set-up, re-wiring and disposing of the ballast. Assume the labor and disposal costs would be $15/ballast. If there are 5,000 ballasts, then at the 12th year, the labor cost to replace the ballasts would be:

($15 in labor and disposal costs/ballast)(5,000 ballasts) = $75,000

The annualized ballast replacement labor cost would be:

($75,000)(1/12 years) = $6,250/year

If the lights are only on 3,750 hours/year (a 25% reduction), the ballasts should last 16 years. This would reduce the annualized ballast replacement labor cost to:

($75,000)(1/16 years) = $4,688/year

Thus, the annualized ballast replacement labor savings are:

$6,250/year – $4,688/year = $1,562/year

Now, we can use similar calculations to quantify the reduced maintenance labor costs for the lamps:

A typical fluorescent lamp life is 20,000 hours.[26,27] If lights are on 5,000 hours per year, the building would need to replace lamps at the 4th year. If there are 10,000 lamps, each costing about $5 in labor to re lamp (including disposal expenses), the replacement expense at the 4th year would be:

($5 in labor/lamp)(10,000 lamps) = $50,000 in labor

The annualized re-lamping labor cost would be:

$50,000/4 = $12,500

If the lights are only on 3,750 hours/year, the lamps should last longer (5.3 years), thereby reducing the annualized labor re-lamping cost to:

$50,000/5.3 years = $9,434/year

Thus, the annualized labor savings are:

$12,500 – $9,434/year = $3,066 per year

Therefore, the total annual avoided maintenance labor costs (ballasts and lamps) are:

$1,562/year + $3,066/year = $4,628/year

This same approach could be used to calculate maintenance labor savings values for other ECMs that involve motors, filters, etc.

Benefit #3: Sample Calculation for Avoided Capital Investment:

Although this additional benefit might not apply within our lighting ECM example, the value can be large when it occurs, so we include another example that demonstrates the calculations below.

Assume that a large capital investment is being planned to meet increasing demand with a budgeted cost of $500,000. Assume that via energy conservation (minimizing leaking HVAC ducts or compressed air systems, reduced operating hours, altered thermostat set points, etc.) you reduce demand such that you can avoid this planned capital investment (additional chiller or compressor to satisfy artificial demand). If a chiller would have cost $500,000 to purchase and install, the annual savings would be the interest on that amount until the chiller is actually purchased. Thus, if the cost of capital (interest rate) is 5%, the saving would be (.05) X ($500,000), or $25,000 per year. Note that if the chiller is permanently avoided, the avoided capital cost is $500,000.

It is worth mentioning that 33% of the energy managers surveyed say this benefit occurred, so it can be very real, especially for HVAC and compressed air systems. Also, if the facility grows but the demand drops, the need for additional capital equipment never arises. We think these savings occur often but are seldom expressed.

Benefit #4: Sample Calculation for Avoided Procurement Costs:

Although this additional benefit might not apply within our Lighting ECM example, we include an example below.

Continuing with the previous example from above, which dealt with avoiding a capital investment, if your organization would have spent additional internal/external costs (legal, engineering, permitting, delays, purchasing, administration, etc.) to procure the chiller, then those avoided costs should also be included if applicable. The authors leave it to the reader to estimate the value of this benefit (if at all), because the value is highly dependent on organization-specific characteristics.

Benefit #5: Sample Calculation for
Avoided Purchases of Carbon Offsets:

If your organization has committed to being *carbon-neutral*, or has some sustainability goal that requires the purchase of carbon offsets when energy is consumed, then there will be tangible savings from energy conservation. When we reduce the amount of energy consumed, we avoid purchasing a certain amount of carbon offsets. Alternatively, this benefit could also be expressed in avoided purchases of renewable energy credits (RECs).[28]

Consider the lighting ECM, which reduced energy consumption by 750,000 kWh/year, or 750 MWh/year. This allows our organization to avoid purchasing 750 RECs/year. If RECs cost $10/MWh, the avoided purchases would equal $7,500/year.

Benefit #6: Sample Calculation for Enhanced Image,
Public Relations, or Recognition:

Although this additional benefit might not apply within our lighting ECM example, we include an example that demonstrates the calculations below.

Assume that your organization has received front-page exposure from an article that was written about your success in energy conservation. If this type of regional advertising normally costs $15,000, then you may be able to consider that as a quantifiable benefit.[29] Alternatively, if your organization has won an award and perhaps that award has allowed your organization to attract better employees, have increased sales of a product, etc., then those are tangible benefits that could be quantified.[30] Many of those surveyed are recognized through EPA with an Energy Star designation. This usually is recognized by the press, and facilities that earn this designation often display their Energy Star plaque at the entrance to a building. This has direct value but is difficult to quantify as the range of values can be large. For example, a small survey of press releases (non-front page exposure) revealed a value of $1,800 per press release/article in local papers.

Benefit #7: Sample Calculation for
Reduced Sales Taxes/Environmental Penalties:

Many utilities apply a sales tax as well as an environmental tax or fee to the total energy cost. Together, these extra costs can represent an additional 8-15%.[31] Continuing with the lighting ECM from before, where we saved $75,000/year, the avoided sales and environmental taxes (assume

10%) could have a value worth:

($75,000/year)(0.10) = $7,500/year

As mentioned before, if (total energy bill $)/(total kWh) is used to evaluate average cost per kWh, the value of this benefit is automatically incorporated into the direct energy dollar savings (in this case $75,000). Thus, if you are using average cost per kWh, Benefit #7 should not apply. For some readers, using an average or "blended" cost is simpler than segregating the kW and kWh savings. However, blended rates assume that impact on demand is equal across the board for ECMs, and that may not be accurate. In another survey, we found that most energy conservation projects impact demand, but perhaps not equally.

Benefit #8: Sample Calculation for Improved Building Value:

Although this additional benefit might not apply within our lighting ECM example, we include an example that demonstrates the calculations below.

This additional benefit is only recognized if the building is sold. Similar to any piece of equipment, its value is partially dependent on the O&M expenses (i.g., a hybrid car has less annual gas expense and thus has additional market value). Thus, if a building is showing less O&M expense it will be worth more when it is sold. If an ECM saves $150,000/year, those savings go directly to the bottom line. The building value would increase by a factor of 10 (given a capitalization factor of 10%):

Increased building value = ($150,000)(10) = $1,500,000—a one-time benefit when the building is sold.

Table 9-2 summarizes the total dollar values of the additional benefits estimated from one sample ECM, using Approach #1. It is noted that it would be rare to have ECMs produce tangible value from all additional benefits, so in this example, *we only included values from half of the listed additional benefits.* The approach and calculations for these benefits could be used as a guide to identify the additional benefits of other ECMs that involve HVAC, motors, etc.

Thus, as we have done above, one method for estimating these additional benefits would be to take case-specific numbers for all ECMs and quantify the values using actual local ballast, bulb, labor, costs, etc. This would take time but would yield very accurate numbers. You could also select which of the benefits apply to your organization and add up those values. (The benefits relevant to your facility may be different than Benefits #1, #2, #5 and #7, which were applied in the lighting ECM example).

Table 9-2. Summary of Additional Benefits from a Lighting Energy Conservation Measure

Additional Benefits (most are Annual)	Value Estimates	% Improvement to Energy Savings
1. Reduced Maintenance Material Costs	$3,616	4.8%
2. Reduced Maintenance Labor Costs	$4,628	6.2%
3. Permanently Avoided Capital Investment	Not Applied to this ECM	
4. Avoided Procurement Costs	Not Applied to this ECM	
5. Avoided Purchases of Carbon Offsets	$7,500	10%
6. Enhanced Image, Public Relations or Recognition	Not Applied to this ECM	
7. Reduced Sales Taxes/Environmental Penalties	$7,500	10%
8. Improved Building Value	Not Applied to this ECM	
Total Additional Value from this ECM	**$23,244**	
% Additional Value Improvement Beyond Energy Savings of $75,000/year		*31%*

Another method would be to use the 31% number that we generate above and simply say, "Our experience shows that additional benefits are worth 31% or more of the energy savings." Thus, if we save $100,000 in energy expenses, we recognize additional benefits worth $31,000.

Alternatively, you could also apply a portion of the 31% if that was a more conservative approach based on your facility's conditions, or based on the specific benefits that your facility receives. For example, if your facility does not purchase carbon offsets and uses an average kWh cost, then you might not receive value from Benefits #5 and #7, yet you would still receive additional benefits worth an 11% improvement to energy savings. Conversely, your facility may experience more benefits than in the example using Approach #1. To illustrate this point, avoiding a capital purchase or improving a building's value would provide benefits that would be substantial and could be much higher.

APPROACH #2: How to Calculate Benefits, Given Organization-wide Budget Data

Below is an approach that could be used for individual cases where more detail is desired. Getting this data would not be elementary, but the approach could be more tailored for individual cases and would establish procedures useful for future documentation.

Using a *budget-based calculation procedure,*

additional value of a benefit = (budget) × [(reduction in operating hours)/(original, or "old," operating hours)] where:

- *Budget* is a figure reflecting how much annually is presently spent for a function (budget line item). By working with accounting personnel you obtain a total cost for that function last year or an average of the previous 3 years.
- *Reduction in operating hours* is the number of reduced operating hours per week or month.
- *Original, or "old" operating hours* is the number of original operating hours for that period
- (Reduction in operating hours)/(original or "old" operating hours) is a fraction between 0 and 1 that reflects the % savings.

Below we show sample calculations for only two budget line items, as the remainder of benefit calculations would follow a similar format.

Benefit 1: Sample Calculation for Reduced Maintenance Material Costs using Budget-based Data

By working closely with accounting personnel, you obtain an average annual cost for labor replacing ballast and bulbs, probably by reviewing all maintenance purchase orders. You find that over the last three years, you spent an average of $19,500 per year on lamps and ballasts. Once you begin turning the lights off 25% of the time, the additional value is:

(Budget)[(Reduction in operating hours)/(Original or old operating hours)]
= (Budget)[(% Savings)]
= ($19,500)(.25)
= $4,875 per year in material savings

This approach is tedious in that it requires careful perusal of three years of accounting data. However, it should be easy to codify the data for future years so that this number can be electronically retrieved at the end of each year.

Benefit 2: Sample Calculation for Reduced Maintenance Labor Costs using Budget-based Data

By carefully examining work order records for the last three years, you find you are spending an average of $23,000 per year on labor to re-place lamps and ballasts.[32] Since you are group relamping, this figure is relatively easy to obtain. Once you begin turning the lights off 25% of the time, the additional value is:

(Budget)[(Reduction in operating hours) / (Original or old operating hours)]

= (Budget)[(% Savings)]

= ($23,000)(.25)

= $5,750 per year in labor savings

Again, the work orders could be codified so that this figure can be easily retrieved electronically each year.

Following similar processes as above, you could develop estimates for all the applicable benefits for which you have budget-based data and compile the values to determine the total value of additional benefits. In all cases, we recommend that a monitoring procedure be designed to validate your savings (going forward). This can be done, but it will require some effort and training at the start.

SUPPLEMENTARY BENEFITS

Beyond the benefits estimated in this research, there are many more which may be applicable to facility managers, as shown in Table 9-3.

CONCLUSION

This chapter has presented additional benefits from energy conservation. From the survey, it is clear there is a high probability that facility managers will experience at least some "additional benefits" from energy conservation. To calculate the value of these benefits, two approaches are provided. One calculates the additional benefits as they relate to a specific energy conservation measure. Another alternative is to collect "tailored" budget data from the whole organization and estimate values.

Within an example application, we found that the additional benefits contributed an additional value worth 31% beyond the energy savings per year. **Perhaps you will estimate more (or less) value at your facility—but it is clear that these additional "CORE" benefits exist and are highly probable.**

You may also choose to only apply a portion of the 31% if that is a more conservative approach based on your facility's conditions, or based

Table 9-3. Additional Benefits from a Focused Effort on Energy Conservation

Additional Benefit	Examples
Utility Rate Reduction	By focusing on energy conservation and learning alternative rate schedules, your organization was able to switch to a lower energy rate structure.
Identification and Capture of all Utility and/or Government Rebates	By focusing on energy conservation, your organization was able to acquire rebates.
Recovered Billing Errors	By focusing on energy conservation, your organization received billing credits from the utility. History shows that energy managers often uncover billing errors when schedules and bills are reviewed carefully.
Reduced Risk to Environmental and/or Legal Costs	Since they last longer, there is less cost for disposal/recycling of bulbs, ballasts, motors, etc. and less environmental risk.
Increased Training and Performance of Facility Staff	As your energy saving program matures, so does the staff running the buildings and equipment. Better understanding of the functions yields reduced operating costs in ancillary areas.
Improved Ability to Manage Energy and Assign 3rd Party Costs to the 3rd Party	As a focus of energy is deployed across the organization, management can be more astute in dealing with external contractors, as well as the parasitic energy consumption they create during construction of new facilities.
Improved Compliance with Building Standards	As more attention is paid to comfort and ventilation requirements, there is better compliance with ASHRAE 62, 55, and other standards
Utility Savings Applied to Staff Positions	Utility savings can be used to fund new positions or avoid staff layoffs
Improved Staff Comfort and Productivity	Through optimization, productivity can improve (comfort, outside air, personnel etc)
Water and Sewer Savings	Through optimization, water consumption and sewerage costs are reduced.

on the specific benefits that your facility receives. Conversely, your facility may receive a greater number of benefits than we showed in the example. For example, *avoiding a capital purchase or improving a building's value would provide benefits that would be substantial.*

We hope that this chapter motivates additional action for energy conservation, dollar savings, and environmental benefits. *Please give these approaches a try, and let us know about your results!*

Footnotes

1. Woodroof, E., Turner, W. and Heinz, S. (2008), "The Secret Benefits from Energy Conservation Contribute Value Worth An 18% Improvement To Energy Savings," *Strategic Planning for Energy and the Environment, Vol 28(1).*
2. Komor, P., (1999), "Selling Energy Efficiency: Multiple Benefits, Multiple Messages," *E Source: The Large Commercial Series LC-1 Report — April 1999.*
3. This paper is focused on "additional benefits" that occur as a result of classic energy conservation techniques—primarily turning off equipment when not needed or by optimizing equipment to reduce wasteful losses. These are actions that usually do not require a capital investment. *Additional research would be required to include energy efficiency measures that require capital investments, which do save energy, but may require the systems to be "on" to save energy. Additional examples of such projects include: installing more efficient equipment such as new chillers, different light sources, or any measure that saves energy not through demand-side reductions.*
4. These "CORE Benefits" have also been labeled as "Secret Benefits" as well as other acronyms such as "MAC," which means "Maintenance, Administrative and Capital" Benefits.
5. "Soft" benefits can include items like improved morale. *For Example: some clients that have implemented energy conservation programs have experienced a definite increase in morale when employees feel they are doing their part to help the environment and are empowered to make decisions to conserve energy. Not every employee is motivated by these factors, but for those that are, a strong energy conservation program adds a global aspect to every position that can lead to improved job satisfaction and increased productivity.*
6. Installation of better equipment (supply side) activities will be covered in future research.
7. If replacement occurs at failure or based on run time, these savings automatically occur. If replacements are planned in advance, planners should adjust their schedules to ensure savings are captured from extended equipment lives (not replacing assets prematurely).
8. Note that a facility may still need to meet minimum ventilation requirements.
9. Labor savings are direct savings when using "external" personnel. When man-hour savings exist using "in-house" salaried personnel, the personnel must be re-allocated to other useful purposes, otherwise there are no direct dollar savings (you still are paying the personnel—even if they aren't working as hard).
10. Alternatively viewed, the company can grow a little without additional personnel.
11. If the purchase is delayed, then the value equals the interest expense or opportunity cost on the avoided capital investment.
12. It is also interesting to note that even the voluntary carbon markets for carbon offset projects have grown substantially during 2008-2012, even during a stagnant global economy. Thus, more organizations (public and private) are committing to reducing

carbon or improving their sustainability programs.

13. Some day, avoided carbon emissions may have direct sellable value. In today's conditions, the tangible value exists if you are trying to be "carbon neutral" and these are offsets you don't have to buy.

14. Saving energy reduces pollution (mercury, SO_x, NO_x, greenhouse gases, etc.) from fossil-fueled power sources. Therefore, energy conservation is an effective sustainability strategy. *For Example, Frito Lay's Sun Chips brand has received significant sales increases due to recognition of energy projects completed at its manufacturing facility in Modesto, CA.*

15. You should investigate whether your existing calculations of energy savings include taxes and environmental penalties to avoid "double counting." However, many facility managers and energy engineers do not include the additional taxes and environmental penalties that are assessed on a unit energy basis, as these penalties/fees can be hidden within the bills, or rate structure's fine print.

16. Woodroof, E., (2011) "Sales Taxes and Utility Rebates can Yield Massive Savings," *Buildings.com Magazine*, October 2011 Issue.

17. When evaluating businesses and buildings, one technique is to estimate value as a multiple (7 to 12) of the annual profit. Because energy savings ultimately reduce costs and therefore improve profits, they raise the value of a building at the time of sale.

18. Alternatively, several studies (research from real estate managers and coordinated by the Institute for Market Transformation between 2009-2011) show that the sale price of a building can increase by 2% to 25% after a commercial building has attained the Energy Star label. Occupancy and Rental rates also experienced a premium price after being labeled Energy Star.

19. Wingender, J. and Woodroof, E., (1997) "When Firms Publicize Energy Management Projects: Their Stock Prices Go Up"—How much—21.33% on Average! *Strategic Planning for Energy and the Environment*, Summer Issue 1997.

20. Griffin, P. and Sun, Y., (2012) "Going Green: Market Reaction to CSR Newswire Releases," University of California. Griffin, Paul A. and Sun, Yuan, Going Green: Market Reaction to CSR Newswire Releases (January 29, 2012). Available at SSRN: http://ssrn.com/abstract=1995132 or http://dx.doi.org/10.2139/ssrn.1995132

21. *For Example: an energy conservation program that saves $100,000 in operating costs is equivalent to generating $1,000,000 in new revenue (assuming the organization has a 10% profit margin). It is more difficult to generate $1,000,000 in new revenue, and would require more marketing, infrastructure, etc. Thus, the energy conservation/efficiency program is an investment with less risk and quickly improves cash flow.*

22. The majority of survey respondents were in educational environments, however the results can apply in most facilities that are saving energy.

23. We labeled these as "Supplementary Benefits" as we did not include calculations for them in this paper.

24. The data could be further stratified to further evaluate the responses by facility size, location or other parameters. This information may be used during future research in an attempt to determine the average value of each "additional benefit."

25. If replacement occurs at failure or based on run time, these savings automatically occur. If replacements are planned in advance, planners should adjust their schedules to insure savings are captured from extended equipment lives (not replacing assets pre-maturely).

26. Lamp life is rated at the factory by turning lamps on and off every three hours until they burn out. If the frequency of on/off cycling is less than 3 hours, lamp lives will decline by 25% on average. Therefore, turning a lamp off for longer periods is better than shorter periods. *For example, it is better to find locations where you can turn off lamps for 5 hours out of 15 hours, instead of 1 minute out of every 3 minutes, although the % time off*

is the same.

27. Additional information on emissions, offsets and reporting can be found in this article: Woodroof, E. (2011), "GHG Emissions Management for Dummies," *Strategic Planning for Energy and the Environment*, Vol 31(2)

28. $15,000 is a sample price based on real costs of "cover" advertising in newspapers in the US and Asia, however it is just a sample—thus the reader should apply an estimated price based on appropriate/available advertising within the local geographic area.

29. *Wall Street Journal*, November 13th, 2007—"How Going Green Draws Talent, Cuts Costs."

30. Samples from several energy audits completed by Profitable Green Solutions, LLC.

31. To get this figure, you must peruse work orders to determine the total man-hour dollars spent for replacing lamps and ballasts as before.

Chapter 10

Overcoming the Barriers that Delay "Good" Projects*

Eric A. Woodroof, Ph.D., CEM

ABSTRACT

Although the popularity of energy management and "green" projects is improving, there are *many* good projects that are postponed or cancelled due to common barriers. This article discusses these barriers and problems, as well as effective, proven strategies to overcome them. These timeless, cutting-edge strategies involve marketing, educational resources, and financing approaches to make your projects *irresistible*. The goal of this article is to help organizations get more good energy management/green projects approved and implemented, thereby helping to slow global warming.

INTRODUCTION

The polar ice is melting and, as far as the planet is concerned, engineers are *wasting their time* if the projects they so carefully develop are not implemented and deliver no value. This article refers to "good" projects as those with a three-year payback or less. Why don't good projects get implemented? There are a variety of reasons and a few common barriers:

---

*Published in *Strategic Planning for Energy and the Environment*, 2008

- *Marketing* (under-marketing a project's value)
- *Education and collaboration* (not expanding the value of a project)
- *Money* (not having a positive cash flow solution)

If a project can't satisfy these criteria, it probably won't be implemented anyway… so focus on the ones that will!

PROBLEM #1: MARKETING

People often ask me why marketing is first on the list. Answer: because NOTHING HAPPENS WITHOUT A SALE. For example, your first job (or your first date) began with you "selling yourself." This is the goal of an interview. In fact, the development of every product/service begins with someone selling a solution to some type of problem. I am saying that selling/convincing is neither "bad" nor unethical. Convincing others when it improves their lives is good—it can be done with passion! When something (like an energy management project) is great, we should sell the benefits *with all the passion in the world*. You would do the same when talking to your kids about "getting a good education," or "learning good manners." Passion can also emerge from fear, such as from the chaos and violence that occurs during an electrical blackout. *Most of the time, humans are more passionate and action-oriented when they are at risk of losing something versus the possibility of gaining something.*

So, we must communicate in a way that the audience (the buyer or project endorser) can understand the problem/pain that they are in now. After they agree that they are "in pain," then they will want to hear about potential solutions.

Attention

It starts with getting the buyer's attention on the problem, the pain it is causing, and a sense of urgency to solve the problem. Only then will a solution seem to be logical. In addition, after the problem/pain is understood, they will be able to become passionate about the solution.

If you fail to get the attention of the endorsers, you are actually doing them a disservice; they won't know they are in trouble and are wasting money. It's like allowing someone to bleed to death when they don't even know they are cut. So, don't be shy—you have a duty to perform.

Warning! Some endorser's personalities' won't like to hear about problems/pain. They may "put their head in the sand" like an ostrich when problems are discussed. Don't blame them; it's part of their personality (which has strengths in other areas). Discover ways to communicate in a way that they will respond. FYI, it can take *seven* impressions (explanations/presentations) before some people will agree on the problem and take action on a solution. Don't give up and don't be surprised or depressed when they don't take action after the first impression.

Below are a few examples of effective headlines* that can help get the attention of an endorser. Feel free to use these in executive summaries:

- "How will the shareholders feel about us throwing money away every month?"
- "A way to make money while reducing emissions…"
- "What will we do with the yearly savings?"
- "We are paying for energy-efficiency projects, whether or not we do them!"
- "Guaranteed, high-yield investments…"
- "If you enjoy throwing money away every month, don't read this…"
- "4.6 billion years of reliability… solar energy."
- "This project could improve our stock price by over 20%!"**
- "Good planets are hard to find."

There are many other great proven examples that are available.† However, you can experiment by looking around for "marketing copy" in magazine advertisements, commercials, etc. There is a reason they call it "copy"—some of the principles are thousands of years old, and they still work! Just change the words to relate to your problem/solution. Try a few versions and test, test, test to see which ones are most effective. Go for it!

Benefits*

After you have their attention, be sure that you include compelling benefits that "take away the pain" the audience is feeling. As engineers, we are good at mentioning the typical benefits:

---

*Ultimate Marketing for Engineers Course, www.ProfitableGreenSolutions.com
**Wingender, J. and Woodroof, E., (1997) "When Firms Publicize Energy Management Projects: Their Stock Prices Go Up"—How much—21.33% on Average! *Strategic Planning for Energy and the Environment*, Summer Issue 1997.
†The "Vault Files," www.ProfitableGreenSolutions.com

- Saves energy, money, waste and emissions;
- Offsets the cost of a planned capital project;
- Improves cost-competitiveness, productivity, etc.;
- Is a relatively low-risk, high-profit investment that directly impacts the bottom line**.

In today's green-minded economy, we could also demonstrate that "green" projects are a very effective marketing tool—which could get the client's marketing department behind your project—because these benefits have also been proven[†]:

- Improves the client's "green" image;
- Differentiates the client from the competition[††];
- Introduces them to new markets, suppliers and clients[‡][¶];
- Helps them grow sales/revenue.

However, we should also mention the passionate, global and moral reasons behind a good "green" project:

- Slows global warming, reduces acid rain;
- Reduces mercury pollution, which allows us to eat healthy;
- Improves our national energy independence;
- Reduces security/disaster risks, etc.

Dollar values for these benefits can often be calculated and should be included in your proposal. To calculate the "green benefit equivalencies," such as "number of trees planted" (from reduced power plant emissions), see the "Money" section of this article.

---

*Download the FREE emissions calculator from www.ProfitableGreenSolutions.com

**For Example: an energy-efficient project that saves $100,000 in operating costs is equivalent to generating $1,000,000 in new sales (assuming the company has a 10% profit margin). It can be more difficult to add $1,000,000 in sales, and would require more infrastructure, etc.

[†]Several examples include: Patagonia, Google, GE, Home Depot, etc. Other examples can be downloaded from the "Resource Vault" at www.ProfitableGreenSolutions.com

[††]For Example: A construction firm switched to hybrid vehicles, which offsets the carbon emissions. The firm's name is prominently displayed on each vehicle. They get tons of new business because they are seen and known as the "greenest construction firm" in the city. Plus, they charge a premium for their services!

[‡]For Example: A law firm renovated their office in a "green" manner and attracted a new client (who chose the firm due to its "green" emphasis). The new client was worth an extra $100,000 in revenue in the first month.

[¶]Additional Examples: "Green" networking groups such as "greendrinks.org" can supplement the traditional business of networking clubs like Rotary Club, Kiwanis, Chamber of Commerce, etc. Also, when joining groups such as the Climate Action Registry, companies are exposed to other members, who could be superior suppliers, clients and partners.

The list above can be expanded, refined and optimized for any project. To build a list like the one above, one technique is "WSGAT"—"What is So Good About That?" Ask that question for every project feature and you will develop a long list of passionate benefits. By the way, this approach has been used in TV sales and has helped sell billions of dollars of material*. If they can sell this much junk on TV, we should have no problem selling green projects that are factually saving the planet! Add the emotional benefits of going "green," and you will have a project that touches the hearts of leaders in your organization.

Call to Action

The "call to action" becomes easy and logical when all of the benefits have been quantified and they are aligned with the client's strategic objectives. Tell endorsers what you want them to do and why. Be sure to include the "cost of delay" in your executive summary. Remind them that they are "in pain" and the project/solution will solve it. Visual aids can be helpful. For example, during one presentation, buckets of dollars were shoveled out a window to demonstrate the losses that were occurring every minute. The executives were literally in pain watching those dollars fly away. They couldn't stand it, and they took action. It is OK to get creative and have some fun in your presentation!

But wait... there is more!

"Configuring" your presentation can make the difference between immediate approval and further delay. There are many ways to configure or "package" your product/project so that it is IRRESISTIBLE. One way to do this is to find a way to make a project's performance guaranteed or "risk-free." Another way is to separate (or add) one part of the project and introduce it as a "free bonus." Everyone likes a "FREE" bonus—it helps them understand that they are getting a good deal. For example, on a recent "green," facility-related project, carbon offsets for a company's fleet were included as a free bonus. The bonus delighted the client and distinguished the project (adding extra value), yet the additional costs were less than $1,000.

Engineers can be two, three or ten times more productive by developing sales and marketing skills. However, there is another reason for developing these sales/marketing skills: Your Career! The skills you learn

---

*Marketing to Millions Manual, Bob Circosta Communications, LLC.

will be valuable to your organization (as well as other organizations). These skills are transferable to other industries too. So keep this in mind when you are investing in yourself; there will almost always be a fantastic pay-off.

Finally, there are two prerequisites that a buyer must see in you before any sale is made: "Trust" and "Value." As far as trust goes, it must be earned, and once earned… it must be cherished. To accelerate the buyer's trust in you, be an advocate for the client and put their needs ahead of your own. Assume the role of their "most trusted advisor," and then deliver. Value comes from applying knowledge, tools, resources, partners, etc., in the best way for the client, which is why education and collaboration is such an important component of success. This is discussed further in the next section. Be sure to read the sub-sections on reciprocal business agreements, joint ventures, and incentives/rebates… great ideas!

PROBLEM #2: Education & Collaboration

Knowing how to deliver the value is an area that requires continuous updating. Today, with the proliferation of energy/green technologies, it is impossible for one person to know all the ways to add value to a project. Green specialties are expanding every day, for example: energy efficiency, water efficiency, green janitorial, LEED*, recycling, transportation, etc.

Learn all you can, then collaborate with other professionals who are also actively learning, and the value available to your clients' increases exponentially. It is important to be open to new ideas and fresh perspectives in this process. "Mind-sharing" or brain-storming techniques can facilitate the process and maximize the number of useful ideas.*

Fortunately, education is a low-cost investment. Collaboration and even joint-ventures/partnering can be done inexpensively as well—and the returns can be huge!

Free sources of green/energy efficiency education:
- https://www.aeecenter.org/seminars/
- http://www.eere.energy.gov/
- http://www.ashrae.org/education/

---

*LEED = Leadership in Energy & Environmental Design
*Results from the Profitable Green Strategies Course, www.ProfitableGreenSolutions.com

- www.usgbc.org
- www.ase.org
- www.energystar.org
- http://greeninginterior.doi.gov

In addition, there are many innovative ways to bring more value to a project. Some include:

- Reciprocal business agreements
- Joint ventures
- Free tax and utility incentives/rebates

Reciprocal Business Agreements

For example: After presenting a $1,000,000 service contract for a global car rental company, the deal was sweetened with an agreement on our part to choose that car company while traveling, which generated over $1,000,000 in extra car rentals for them. To the client, they were getting an extra $1,000,000 in revenue by working with us versus the competitors. With suppliers, partners, colleagues, professionals, etc. could you develop reciprocal business agreements? How could you help two clients (or suppliers) benefit from each other? How could you help them become more "green?"

Another example: We helped client #1 supply green solutions to client #2. Both clients were extremely happy to generate more sales/and save money. When it was time to approve our next round of projects, there was little resistance because we had helped them earn/save far more than the costs of the proposed projects. This illustrates the value of being the "trusted advisor."

Joint Ventures

For example: A "green" travel agency gives 50% of its commissions back to its clients in exchange for their travel business.* The client can use this extra, free money to fund "green" initiatives or scholarships, or other social programs. The travel agency guarantees the lowest prices and easily doubles its business because it delivers more value to its clients via joint ventures.

---

* www.GreenTravelPartners.com

Free Tax and Utility Incentives/Rebates

For example: In California, 50% of a solar project was funded by federal and state rebates. Utility incentives lowered the installation costs even further. There are numerous free tax and utility incentives available, and some are discussed in the next section.

In addition to the options above, many utilities and third parties are offering "green power purchase agreements," which are essentially "wind and solar performance contracts." For example, if you want to put solar panels on your roof, a third party (often a utility or solar contractor) finances the project installation and then sells you the renewable energy produced from your roof (at a known price) for 15-25 years. So you get "green" power at no up-front cost, as well as a known future energy cost (lowers your risk to energy price volatility). The financier wins because the project will pay back the investment within 10 years and the rest is profit.

There are an unlimited number of creative "win-win" contracts available. However, before finalizing or even developing your solution, be sure that you understand the client's strategic and financial goals, then align the value to support the client's larger objectives.

PROBLEM #3: MONEY

If you do a good job tapping into the passion behind the project and are satisfying the emotional, financial and other approval criteria, you should have enough benefits to get the project approved, especially if the project is above the client's MARR.* However, if your organization is capitally constrained, you can finance a project and have positive cash flow. CFOs like positive cash flow projects! Cash flow constraints (not having the up-front capital to install a project) represent over 35% of the reasons why projects are not implemented[†].

Financing does not have to be complicated. In fact, financing energy efficiency/green projects can be very similar to your mortgage or car payment—fixed payments for a length of time. However, with a

---

*MARR= Minimum Attractive Rate of Return. For more info on this topic see: Woodroof, E., Thumann, A. (2005) *Handbook for Financing Energy Projects*, Fairmont Press, Atlanta.
[†]U.S. Department of Energy, (1996) "Analysis of Energy-Efficiency Investment Decisions by Small and Medium-Sized Manufacturers," U.S. DOE, Office of Policy and Office of Energy Efficiency and Renewable Energy, pp. 37-38.

good project, you can finance the project such that the annual savings are greater than the finance payments, which means the project becomes "cash flow positive" and does not impact the capital budget! This can allow the endorser to move forward without sacrificing any other budget line item.

Table 10-1 shows the cash flow for a non-financed project*. Assume the project costs $100,000 and saves $28,000 per year for 15 years. This project could get approved IF the client has $100,000 in cash to fund it. The project has a net present value of $ 102,700 and an internal rate of return of 27%.

Now, let's look at financing the project with a simple loan. Let's say the client finances the $100,000 for 15 years at 10% per year. That means that instead of investing $100,000 up front (the bank provides these funds), the client pays $13,147 each year to the bank for 15 years. At the end of 15 years, the bank loan is paid off (just like a mortgage or car payment—just a different time period). To keep this simple, ignore interest

Table 10-1. Project Cash Flow (paid with cash)

EOY	Savings	Cost	Cash Flow
0	-	(100,000)	$ (100,000)
1	28,000		$ 28,000
2	28,000		$ 28,000
3	28,000		$ 28,000
4	28,000		$ 28,000
5	28,000		$ 28,000
6	28,000		$ 28,000
7	28,000		$ 28,000
8	28,000		$ 28,000
9	28,000		$ 28,000
10	28,000		$ 28,000
11	28,000		$ 28,000
12	28,000		$ 28,000
13	28,000		$ 28,000
14	28,000		$ 28,000
15	28,000		$ 28,000
NPV i=10%			$102,700
IRR			27%

*Advanced Project Financing Course, www.ProfitableGreenSolutions.com

tax deductions as well as depreciation, which would likely improve the financials even further.

Table 10-2. Financed Project Cash Flow

EOY	Savings	Finance Cost	Cash Flow
0	-	-	$ -
1	28,000	13,147	$ 14,853
2	28,000	13,147	$ 14,853
3	28,000	13,147	$ 14,853
4	28,000	13,147	$ 14,853
5	28,000	13,147	$ 14,853
6	28,000	13,147	$ 14,853
7	28,000	13,147	$ 14,853
8	28,000	13,147	$ 14,853
9	28,000	13,147	$ 14,853
10	28,000	13,147	$ 14,853
11	28,000	13,147	$ 14,853
12	28,000	13,147	$ 14,853
13	28,000	13,147	$ 14,853
14	28,000	13,147	$ 14,853
15	28,000	13,147	$ 14,853
NPV i=10%			$112,970
IRR			n/a

In this case, the project generates $14,853 each year for the client. Because there is no up-front investment required, the IRR value becomes infinity.

Adding in the "Green Benefits" could further illustrate the project's benefits. Table 10-3 shows what some of these benefits could include. Note that it can be easier for the audience to visualize equivalencies ("car miles not driven," or "trees planted") instead of lbs of CO_2.

However, there are even more benefits when you consider the following impacts the project could have on:

- Shareholders in the annual report,
- Community morale and "green image,"
- Productivity improvements,
- Legal risk reduction,

Table 10-3. Green Benefits*

kWh Saved per Year	260,000
# of Years	15
kWh Saved during Project	3,900,000

Barrels of Oil Not Consumed	7,917	
Car Miles Not Driven	6,924,450	
Acid Rain Emission Reduction	29,250	lbs of Sox
Smog Emission Reductions	14,040	lbs of Nox
GreenHouse Gas Reduction	6,045,000	lbs of CO2
Mature Trees Planted	13,260	

- LEED points, white certificates, RECs[†],
- FREE public press[¶].

FREE Money

In addition, there are utility rebates, tax refunds, credits, and other sources of free money that will improve a project's financial return. Here are some useful web sites that allow you to see utility and tax benefits in your state:

- www.dsireusa.org/
- www.energytaxincentives.org
- http://www.efficientbuildings.org
- http://www.lightingtaxdeduction.org/

But don't just rely on web sites. Use professionals; they should know what techniques, technologies, and rebates are best for your geographic area.

SUMMARY

This article has described the three common barriers (marketing, education, and money), as well as provided a start on how to overcome them. To get a project approved:

---

*Download the FREE emissions calculator from www.ProfitableGreenSolutions.com
[†]REC = Renewable energy credit
[¶]Press release samples from the "Vault" at www.ProfitableGreenSolutions.com

- Articulate the problem/pain.

- Collaborate to add value in the solution.

- Quantify all the benefits.

- Minimize financial risk.

- Develop/configure an executive summary that "sings" to the heart of the endorser.

Hopefully, these techniques will help you get your next project approved. Why is this important? Because the ice is melting! We are counting on you.

Chapter 11

Opportunities via Codes, Standards and Legislation

Millard Carr

This chapter presents a historical perspective on federal legislation executive orders, key codes, standards, and regulations, all of which have impacted energy policy and are still playing a major role in shaping energy usage. The context of past standards and legislation must be understood in order to properly implement the proper systems and to be able to impact future codes. The Energy Policy Act, for example, has created an environment for retail competition. Electric utilities will drastically change the way they operate in order to provide power and lower cost. This in turn could drastically reduce utility-sponsored incentive and rebate programs, which have influenced energy conservation adoption. The chapter attempts to cover a majority of the material that currently impacts the energy related industries, with respect to their initial writing.

The main difference between standards, codes and regulations is an increasing level of enforceability of the various design parameters. A group of interested parties (vendors, trade organizations, engineers, designers, citizens, etc.) may develop a standard in order to assure minimum levels of performance. The standard acts as a suggestion to those parties involved, but is not enforceable until it is codified by a governing body (local or state agency) that makes the standard a code. Not meeting this code may prevent continuance of a building permit or the ultimate stoppage of work. Once the federal government makes the code part of the federal code, it becomes a regulation. Often this progression involves equipment development and commercialization prior to codification in order to assure that the standards are attainable.

Energy related executive orders direct federal agencies to take specific action to implement presidential energy efficiency and renewable energy application directions in all federal facilities. While these actions are not incumbent on the private sector, they provide a tangible indication of cost effective actions and technology innovation. They also provide strong market opportunities and direction for the private sector to assist federal facilities in responding to the requirements. Energy related executive orders have been signed by every President since Richard Nixon with the exception of Gerald Ford.

Following is a summary of pertinent executive orders and laws implementing energy efficiency improvements in the United States.

EXECUTIVE ORDER 13514

EXECUTIVE ORDER 13514, Federal Leadership in Environmental, Energy, and Economic Performance, signed on October 8, 2009 expanded upon the energy reduction and environmental performance requirements of E.O. 13423. Specifically, it required federal agency heads to designate a senior management official to serve as senior sustainability officer to establish agency green house gas reduction goals and be accountable for agency conformance. It required federal agencies to develop, implement, and annually update a strategic sustainability performance plan that prioritizes agency actions based on life-cycle return on investment. Between fiscal years 2011 and 2021, the order would establish and report to the CEQ chair and OMB director a fiscal year 2020 percentage reduction target of agency-wide scope 1 and scope 2 GHG emissions in absolute terms relative to a fiscal year 2008 baseline through:

- Reducing agency building energy intensity
- Increasing agency renewable energy use and on-site projects
- Reducing agency use of fossil fuels by using low GHG emitting fuels and alternative fuel vehicles, optimizing vehicle numbers across agency fleets and reducing petroleum consumption in agency fleets of 20 or more 2% annually through fiscal year 2020 relative to fiscal year 2005 baseline.

The order would implement high performance sustainable federal building design, construction, operation and management, maintenance, and deconstruction by:

- Ensuring all new federal buildings, entering the design phase in 2020 or later, are designed to achieve zero net energy by 2030.

- Ensuring all new construction, major renovations, or repair or alteration of federal buildings comply with the Guiding Principles of Federal Leadership in High Performance and Sustainable Buildings.

- Ensuring at least 15% of existing agency buildings and leases (above 5,000 gross square feet) meet the Guiding Principles by fiscal year 2015 and that the agency makes annual progress towards 100% compliance across its building inventory.

- Pursuing cost-effective, innovative strategies (e.g., highly-reflective and vegetated roofs) to minimize consumption of energy, water, and materials.

- Managing existing building systems to reduce the consumption of energy, water, and materials, and identifying alternatives to renovation that reduce existing asset deferred maintenance costs.

The order would improve water efficiency and management by:

- Reducing potable water consumption intensity 2% annually through fiscal year 2020, or 26% by the end of fiscal year 2020, relative to a fiscal year 2007 baseline,

- Reducing agency industrial, landscaping, and agricultural water consumption 2% annually, or 20% by the end of fiscal year 2020, relative to a fiscal year 2010 baseline and

- Identifying, promoting, and implementing water reuse strategies

THE AMERICAN RECOVERY AND REINVESTMENT ACT (ARRA) Public Law 111-5

The American Recovery and Reinvestment Act (ARRA PL111-5) signed into law on February 17, 2009 provides numerous energy related tax incentives for businesses:

- Extension of Renewable Energy Production Tax Credit (Section

1101): The new law generally extends the "eligibility dates" of a tax credit for facilities producing electricity from wind, closed-loop biomass, open-loop biomass, geothermal energy, municipal solid waste, qualified hydropower and marine and hydrokinetic renewable energy. The new law extends the "placed in service date" for wind facilities to Dec. 31, 2012. For the other facilities, the placed-in-service date was extended from December 31, 2010 (December 31, 2011 in the case of marine and hydrokinetic renewable energy facilities) to Dec. 31, 2013.

- Election of Investment Credit in Lieu of Production Credit (Section 1102): Businesses who place in service facilities that produce electricity from wind and some other renewable resources after Dec 31, 2008 can choose either the energy investment tax credit, which generally provides a 30 percent tax credit for investments in energy projects or the production tax credit, which can provide a credit of up to 2.1 cents per kilowatt-hour for electricity produced from renewable sources. A business may not claim both credits for the same facility.

- Repeal of Certain Limits on Business Credits for Renewable Energy Property (Section 1103): The new law repeals the $4,000 limit on the 30 percent tax credit for small wind energy property and the limitation on property financed by subsidized energy financing. The repeal applies to property placed in service after Dec. 31, 2008.

- Coordination With Renewable Energy Grants (Section 1104): Business taxpayers also can apply for a grant instead of claiming either the energy investment tax credit or the renewable energy production tax credit for property placed in service in 2009 or 2010. In some cases, if construction begins in 2009 or 2010, the grant can be claimed for energy investment credit property placed in service through 2016, and for qualified renewable energy facilities, the grant is 30 percent of the investment in the facility and the property must be placed in service before 2014 (2013 for wind facilities).

- New Clean Renewable Energy Bonds (Section 1111): The new law increases the amount of funds available to issue new clean renewable energy bonds from the one-time national limit of $800 million to $2.4 billion. These qualified tax credit bonds can be issued to finance certain types of facilities that generate electricity from renewable sources (for example, wind and solar).

- Qualified Energy Conservation Bonds (Section 1112): The new law increases the amount of funds available to issue qualified energy conservation bonds from the one-time national limit of $800 million to $3.2 billion. These qualified tax credit bonds can be issued to finance governmental programs to reduce greenhouse gas emissions and other conservation purposes.

- Temporary Increase in Credit for Alternative Fuel Vehicle Refueling Property (Section 1123): The new law modifies the credit rate and limit amounts for property placed in service in 2009 and 2010. Qualified property (other than property relating to hydrogen) is now eligible for a 50 percent credit, and the per-location limit increases to $50,000 for business property (increases to $2,000 for other/residential locations). Property relating to hydrogen keeps the 30 percent rate as before, but the per-business location limit rises to $200,000.

EXECUTIVE ORDER 13423

Executive Order (E.O.) 13423, *Strengthening Federal Environmental, Energy, and Transportation Management,* signed on January 24, 2007, strengthened key energy program goals for the federal government, set more challenging goals than the Energy Policy Act of 2005 (EPAct 2005) and superseded E.O. 13123 and E.O. 13149.

E.O. 13423 sets numerous federal energy and environmental management requirements in several areas, including but not limited to:

- Implementing Instructions
- Reducing Energy Intensity
- Increasing Use of Renewable Energy
- Reducing Water Intensity
- Designing and Operating Sustainable Buildings
- Managing Federal Fleets

The summary below is intended as a reference only. You should refer to the full text of E.O. 13423 for more details relevant to your work.

Reducing Energy Intensity

E.O. 13423 requires federal agencies to reduce energy intensity by 3% each year, leading to 30% by the end of fiscal year (FY) 2015 compared to an FY 2003 baseline. (This goal was given the weight of law when ratified by EISA 2007.)

Increasing Use of Renewable Energy

To comply with E.O. 13423, federal agencies must ensure that at least half of all renewable energy required under EPAct 2005 comes from new renewable sources (developed after January 1, 1999).

To the maximum extent possible, renewable energy generation projects should be implemented on agency property for agency use. Agencies can also purchase renewable energy to help meet E.O. 13423 requirements.

Reducing Water Intensity

E.O. 13423 mandates that federal agencies reduce water intensity (gallons per square foot) by 2% each year through FY 2015 for a total of 16% based on water consumption in FY 2007.

Designing and Operating Sustainable Buildings

E.O. 13423 requires federal agencies to ensure new construction and major renovations comply with the 2006 Federal Leadership in High Performance and Sustainable Buildings Memorandum of Understanding (MOU), which was signed at the White House Summit on Federal Sustainable Buildings. It also requires that 15% of the existing federal capital asset building inventory of each agency incorporate the sustainable practices in the Guiding Principles by the end of fiscal year 2015.

Managing Federal Fleets

Federal agencies with 20 or more vehicles must ensure fleet petroleum reduction of 2% each year measured against baselines set in fiscal year 2005. Federal agencies must also increase use of alternative fuels by 10% each year compared to the previous year. E.O. 13423 also covers the use of plug-in hybrids.

THE ENERGY INDEPENDENCE AND
SECURITY ACT OF 2007 (H.R.6)

The Energy Independence and Security Act of 2007 (Public Law 110-140) was signed into law December 19, 2007. Key provisions of the law are summarized below.

Title I Energy Security through
Improved Vehicle Fuel Economy

- The Corporate Average Fuel Economy (CAFE) sets a target of 35 miles per gallon for the combined fleet of cars and light trucks by 2020.
- The law establishes a loan guarantee program for advanced battery development, a grant program for plug-in hybrid vehicles, incentives for purchasing heavy-duty hybrid vehicles for fleets, and credits for various electric vehicles.

Title II Energy Security through
Increased Production of Biofuels

- The law increases the Renewable Fuels Standard (RFS), which sets annual requirements for the quantity of renewable fuels produced and used in motor vehicles. RFS requires 9 billion gallons of renewable fuels in 2008, increasing to 36 billion gallons in 2022.

Title III Energy Savings Through
Improved Standards for Appliances and Lighting

- The law establishes new efficiency standards for motors, external power supplies, residential clothes washers, dishwashers, dehumidifiers, refrigerators, refrigerator freezers, and residential boilers.
- It contains a set of national standards for light bulbs. The first part of the standard would increase energy efficiency of light bulbs 30% and phase out most common types of incandescent light bulb by 2012-2014.
- It requires the federal government to substitute energy efficient lighting for incandescent bulbs.

Title IV Energy Savings in Buildings and Industry

- Increases funding for the Department of Energy's Weatherization Program, providing 3.75 billion dollars over five years.
- Encourages the development of more energy efficient "green" commercial buildings, creating an Office of Commercial High Performance Green Buildings at the Department of Energy.
- Sets a national goal to achieve zero-net energy use for new commercial buildings built after 2025. A further goal is to retrofit all pre-construction 2025 buildings to zero-net energy by 2050.
- Requires that total energy use in federal buildings (relative to the 2005 level) be reduced 30% by 2015.
  - For those new buildings and retrofit projects costing over $2.5 million—fossil generated energy use relative to 2003 must be reduced 50% by 2010, 65% by 2015, 80% by 2020, 90% by 2025 and 100% by 2030.
- Requires federal facilities to conduct a comprehensive energy and water evaluation for each facility at least once every four years. Evaluations to include the need for re- and retro-commissioning and the requirement to implement all life cycle cost effective projects found within two years of the evaluation.
- Requires the addition of steam and natural gas meters to the electric metering requirement of EPACT 05.
- Requires that not less than 30% of hot water demand for new Federal buildings be from solar hot water heaters if life cycle cost effective.
- Requires that all federally purchased electrical equipment must have standby power requirements of less than one watt.
- Eliminated the sunset on the authority for federal facilities to use Energy Savings Performance Contracts, allowed sales of excess energy form cogeneration or remewable energy sources and restricted the agencies from limiting such contracts to less than 25 years or the obligation of them. It further requires training of contract officers in the negotiation of alternatively financed energy contracts.
- Requires in 2012 and each year thereafter federal fleets will decrease use of petroleum by at least 20% and increase use of alternative fuels by 10% if such fuel can be certified to result in less greenhouse gas emissions.

- Requires at least one renewable fuel pump be installed at each federal fleet fueling center of over 100,000 gallons/year by November 2011.
- Requires that each federal agency ensure that major replacements of installed equipment (such as heating and cooling systems) or renovation/expansion of existing space employ the most energy efficient designs, systems, equipment, and controls that are life cycle cost effective. For the purpose of calculating life-cycle cost calculations, the time period will increase from 25 years in the prior law to 40 years.
- Directs the Department of Energy to conduct research to develop and demonstrate new process technologies and operating practices to significantly improve the energy efficiency of equipment and processes used by energy-intensive industries.
- Directs the Environmental Protection Agency to establish a recoverable waste energy inventory program. The program must include an ongoing survey of all major industry and large commercial combustion services in the United States.
- Includes new incentives to promote new industrial energy efficiency through the conversion of waste heat into electricity.
- Creates a grant program for Healthy High Performance Schools that aims to encourage states, local governments and school systems to build green schools.
- Creates a program of grants and loans to support energy efficiency and energy sustainability projects at public institutions.

Title V Energy Savings in Government and Public Institutions

- Promotes energy savings performance contracting in the federal government and provides flexible financing and training of federal contract officers.
- Promotes the purchase of energy efficient products and procurement of alternative fuels with lower carbon emissions for the federal government.
- Reauthorizes state energy grants for renewable energy and energy efficiency technologies through 2012.
- Establishes an energy and environmental block grant program to be used for seed money for innovative local best practices.

Title VI Alternative Research and Development

- Authorizes research and development to expand the use of geo-thermal energy.
- Improves the cost and effectiveness of thermal energy storage technologies that could improve the operation of concentrating solar power electric generation plants.
- Promotes research and development of technologies that produce electricity from waves, tides, currents, and ocean thermal differences.
- Authorizes a development program on energy storage systems for electric drive vehicles, stationary applications, and electricity transmission and distribution.

Title VII Carbon Capture and Sequestration

- Provides grants to demonstrate technologies to capture carbon dioxide from industrial sources.
- Authorizes a nationwide assessment of geological formations capable of sequestering carbon dioxide underground.

Title VIII Improved Management of Energy Policy

- Creates a 50% matching grants program for constructing small renewable energy projects that will have an electrical generation capacity less than 15 megawatts.
- Prohibits crude oil and petroleum product wholesalers from using any technique to manipulate the market or provide false information.

Title IX International Energy Programs

- Promotes US exports in clean, efficient technologies to India, China and other developing countries.
- Authorizes US Agency for International Development (USAID) to increase funding to promote clean energy technologies in developing countries.

Title X Green Jobs

- Creates an energy efficiency and renewable energy worker training program for "green collar" jobs.
- Provides training opportunities in the energy field for individuals who need to update their skills.

Title XI Energy Transportation and Infrastructure

- Establishes an office of climate change and environment to coordinate and implement strategies to reduce transportation related energy use.

Title XII Small Business Energy Programs

- Loans, grants and debentures are established to help small businesses develop, invest in, and purchase energy efficient equipment and technologies.

Title XIII Smart Grid

- Promotes a "smart electric grid" to modernize and strengthen the reliability and energy efficiency of the electricity supply. The term "Smart Grid" refers to a distribution system that allows for flow of information from a customer's meter in two directions, both inside the house to thermostats, appliances, and other devices; and from the house back to the utility.

NATIONAL DEFENSE AUTHORIZATION ACT (NDAA 2007)

(Public Law 109-364) signed October 17, 2006 provides specific congressional direction for energy program activities in the Department of Defense.

Section 2852

- Establishes as a DOD goal to: (1) produce or procure not less than 25% of the total electric energy it consumes during FY2025 and thereafter from renewable energy sources; and (2) produce or procure such renewable energy when it is life-cycle cost-effective to do so.

Section 2911
- Directs the development and reporting of energy performance goals and a plan for achieving them.

Section 2912
- Authorizes the availability and use of energy cost savings.

Section 2913
- Directs the establishment of a simplified process to enter into Shared Energy Savings Contracts and encourages Defense facilities to enter into energy savings contracts with gas and electric companies.

THE ENERGY POLICY ACT OF 2005

The first major piece of national energy legislation since the Energy Policy Act of 1992, EPAct 2005 was signed by President George W. Bush on August 8, 2005 and became effective January 1, 2006. The major thrust of EPAct 2005 is energy production. However, there are many important sections of the act that do help promote energy efficiency and energy conservation. There are also some significant impacts on federal energy management. Highlights are described below:

Federal Energy Management
- The United States is the single largest energy user, with about a $10 billion energy budget. Forty-four percent of this budget was used for non-mobile buildings and facilities. The United States is also the single largest product purchaser, with $6 billion spent for energy using products, vehicles, and equipment.

Energy Management Goals
- An annual energy reduction goal of 2% is in place from fiscal year 2006 to fiscal year 2015, for a total energy reduction of 20%.

- Electric metering is required in all federal building by the year 2012.

- Energy efficient specifications are required in procurement bids and evaluations.

- Energy efficient products to be listed in federal catalogs include

Energy Star and FEMP recommended products by GSA and Defense Logistics Agency.

• Energy Service Performance Contracts (ESPCs) are reauthorized through September 30, 2016.

• New federal buildings are required to be designed 30% below ASHRAE standard or the International Energy Code (if life-cycle cost effective.) Agencies must identify those that meet or exceed the standard.

• Renewable electricity consumption by the federal government cannot be less than: 3% from fiscal year 2007-2009, 5% from fiscal year 2010-2012, and 7.5% from fiscal year 2013-present. Double credits are earned for renewables produced on the site or on federal lands and used at a federal facility, as well as renewables produced on Native American lands.

• The goal for photovoltaic energy is to have 20,000 solar energy systems installed in federal buildings by the year 2012.

Tax Provisions

• Tax credits will be issued for residential solar photovoltaic and hot water heating systems. Tax deductions will be offered for highly efficient commercial buildings and highly efficient new homes. There will also be tax credits for improvements made to existing homes, including high efficiency HVAC systems and residential fuel cell systems. Tax credits are also available for fuel cells and microturbines used in businesses.

THE ENERGY POLICY ACT OF 1992

The Energy Policy Act of 1992 is far-reaching, and its implementation is impacting electric power deregulation, building codes, and new energy efficient products. Sometimes policy makers do not see the extensive impact of their legislation. This comprehensive legislation significantly impacts energy conservation, power generation, and alternative fuel vehicles, as well as energy production. The federal and private sectors are impacted by this comprehensive energy act. Highlights are described below:

Energy Efficiency Provisions
Buildings
*   Requires states to establish minimum commercial building energy codes and to consider minimum residential codes based on current voluntary codes.

Utilities
*   Requires states to consider new regulatory standards that would require utilities to undertake integrated resource planning, allow efficiency programs to be at least as profitable as new supply options, and encourage improvements in supply system efficiency.

Equipment Standards
*   Establishes efficiency standards for commercial heating and air-conditioning equipment, electric motors, and lamps.
*   Gives the private sector an opportunity to establish voluntary efficiency information/labeling programs for windows, office equipment and luminaries (or the Department of Energy will establish such programs).

Renewable Energy
*   Establishes a program for providing federal support on a competitive basis for renewable energy technologies. Expands program to promote export of these renewable energy technologies to emerging markets in developing countries.

Alternative Fuels
*   Gives Department of Energy authority to require a private and municipal alternative fuel fleet program, starting in 1998. Provides a federal alternative fuel fleet program with phased-in acquisition schedule; also provides state fleet program for large fleets in large cities.

Electric Vehicles
    Establishes comprehensive program for the research and development, infrastructure promotion, and vehicle demonstration for electric motor vehicle.

Electricity
*   Removes obstacles to wholesale power competition in the Public Utilities Holding Company Act by allowing both utilities and non-

utilities to form exempt wholesale generators without triggering the PUHCA restrictions.

Global Climate Change

- Directs the Energy Information Administration to establish a baseline inventory of greenhouse gas emissions and establishes a program for the voluntary reporting of those emissions. Directs the Department of Energy to prepare a report analyzing the strategies for mitigating global climate change and to develop a least-cost energy strategy for reducing the generation of greenhouse gases.

Research and Development

- Directs Dept. of Energy to undertake research and development on a wide range of energy technologies, including energy efficiency technologies, natural gas end-use products, renewable energy resources, heating and cooling products, and electric vehicles.

CODES AND STANDARDS

Energy codes specify how buildings *must* be constructed or perform and are written in a mandatory, enforceable language. State and local governments adopt and enforce energy codes for their jurisdictions. Energy standards describe how buildings *should* be constructed to save energy cost effectively. They are published by national organizations such as the American Society of Heating, Refrigerating, and Air Conditioning Engineers (ASHRAE). Such standards are not mandatory but serve as national recommendations, with some variation for regional climate. State and local governments frequently use energy standards as the technical basis for developing their energy codes. Some energy standards are written in a mandatory, enforceable language, making it easy for jurisdictions to incorporate the provisions of the energy standards directly into their laws or regulations. The requirement for the federal sector to use ASHRAE 90.1 and 90.2 as mandatory standards for all new federal buildings is specified in the Code of Federal Regulations (10 CFR 435).

Most states use the ASHRAE 90 standard as their basis for the energy component of their building codes. ASHRAE 90.1 is used for commercial buildings, while ASHRAE 90.2 is used for residential buildings.

Some states have quite comprehensive building codes (for example, California Title 24).

ASHRAE Standard 90.1

•   Sets minimum requirements for the energy efficient design of new buildings so they may be constructed, operated, and maintained in a manner that minimizes the use of energy without constraining the building function and productivity of the occupants.

•   Addresses building components and systems that affect energy usage.

•   Sections 5-10 are the technical sections specifically addressing components of the building envelope, HVAC systems and equipment, service water heating, power, lighting, and motors. Each technical section contains general requirements and mandatory provisions. Some sections also include prescriptive and performance requirements.

ASHRAE Standard 90.2

•   Sets minimum requirements for the energy efficient design for new low-rise residential buildings.

When the Department of Energy determines that a revision would improve energy efficiency, each state has two years to review the energy provisions of its residential or commercial building code. For residential buildings, a state has the option of revising its residential code to meet or to exceed the residential portion of ASHRAE 90.2. For commercial buildings, a state is required to update its commercial code to meet or exceed the provision of ASHRAE 90.1.

ASHRAE standards 90.1 and 90.2 are developed and revised through voluntary consensus and public hearing processes that are critical to widespread support for their adoption. Both standards are continually maintained by separate Standing Standards Projects Committees. Committee membership varies from ten to sixty voting members. Committee membership includes representatives from many groups to ensure balance among all interest categories. After the committee proposes revisions, the standard undergoes public review and comment. When a majority of the parties substantially agree, the revised standard is submitted to the ASHRAE Board of Directors. This entire process can take anywhere from two to ten years to complete. ASHRAE Standards 90.1 and 90.2 are automatically revised and published every three years. Approved interim revisions are posted on the ASHRAE website (www.ashrae.org) and are included in the next published version.

The energy cost budget method permits tradeoffs between building systems (lighting and fenestration, for example) if the annual energy cost estimated for the proposed design does not exceed the annual energy cost of a base design that fulfills the prescriptive requirements. Using the energy cost budget method approach requires simulation software that can analyze energy consumption in buildings and model the energy features in the proposed design. ASHRAE 90.1 sets minimum requirements for the simulation software; suitable programs include BLAST, eQUEST, and TRACE.

CLIMATE CHANGE

Kyoto Protocol

The goal of the Kyoto Protocol is to stabilize green house gases in the atmosphere that would prevent human impact on global climate change. The nations that signed this treaty come together to make decisions at meetings called "Conferences of the Parties." The 38 parties are grouped into two groups, developed industrialized nations and developing countries. The Kyoto Protocol, an international agreement reached in Kyoto in 1997 by the third Conference of the Parties (COP-3), aims to lower emissions from two groups of three green house gases: (1) carbon dioxide, methane, and nitrous oxide; and (2) hyrdofluorocarbon (HFC): sulfur hexafluoride and perfluorocarbons.

INDOOR AIR QUALITY STANDARDS

Indoor air quality (IAQ) is an emerging issue of concern to building managers, operators, and designers. Recent research has shown that indoor air is often less clean than outdoor air and federal legislation has been proposed to establish programs to deal with this issue on a national level. This, like the asbestos issue, will have an impact on building design and operations. Americans today spend long hours inside buildings, so building operators, managers, and designers must be aware of potential IAQ problems and how they can be avoided.

IAQ problems, sometimes termed "sick building syndrome," have become an acknowledged health and comfort problem. Buildings are characterized as sick when occupants complain of acute symptoms such as headache; eye, nose, and throat irritations; dizziness; nausea; sensitivity to odors; and difficulty in concentrating. The complaints may become more clinically defined so that an occupant may be said to have developed an illness believed to be related to IAQ problems.

The most effective means to deal with an IAQ problem is to remove or minimize the pollutant source, when feasible. If not, dilution and filtration may be effective.

The purpose of ASHRAE Standard 62 is to specify minimum ventilation rates and indoor air quality that will be acceptable to human occupants and thereby minimize the potential for adverse health effects. ASHRAE defines acceptable indoor air quality as the air in which there are no known contaminants at harmful concentrations (as determined by cognizant authorities) and with which a substantial majority of those exposed (80% or more) do not express dissatisfaction.

ASHRAE Standard 55, which sets environmental conditions for human occupancy, covers several environmental parameters including temperature, radiation, humidity, and air movement. The standard specifies conditions in which 80% of the occupants will find the environment thermally acceptable. This applies to healthy people in normal indoor environments for winter and summer conditions. Adjustment factors are described for various activity levels and clothing levels.

The International Performance Measurement and Verification Protocol (IPMVP) is used for commercial and industrial facility operators. It offers standards for measurement and verification of energy and water efficiency projects. The IPMVP volumes are used to: (1) Develop a measurement and verification strategy and plan for quantifying energy and water savings in retrofits and new construction, (2) Monitor indoor environmental quality, and (3) Quantify emissions reduction. (www. evo-world.org)

REGULATORY AND LEGISLATIVE ISSUES IMPACTING COGENERATION AND INDEPENDENT POWER PRODUCTION

Public Utilities Regulatory Policies Act (PURPA)

This legislation was part of the 1978 National Energy Act and has had perhaps the most significant effect on the development of cogeneration and other forms of alternative energy production in the past decade. Certain provisions of PURPA also apply to the exchange of electric power between utilities and cogenerators. It provides a number of benefits to those cogenerators who can become qualifying facilities (QFs) under this act. Specifically, it:

- Requires utilities to purchase the power made available by co-generations at reasonable buy-back rates. These rates are typically based on the utilities' cost.

- Guarantees the cogeneration or small power producer interconnection with the electric grid and the availability of backup service from the utility.

- Dictates that supplemental power requirements of cogeneration must be provided at a reasonable cost.

- Exempts cogenerations and small power producers from federal and state utility regulations and associated reporting requirements of these bodies.

In order to assure a facility the benefits of PURPA, a cogenerator must become a qualifying facility. To achieve this status, a cogenerator must generate electricity and useful thermal energy from a single fuel source. In addition, a cogeneration facility must be less than 50% owned by an electric utility or an electric utility holding company. Finally, the plant must meet the minimum annual operating efficiency standard established by the Federal Energy Regulatory Commission (FERC) when using oil or natural gas as the principal fuel source. The standard is that the useful electric power output, plus one half of the useful thermal output of the facility, must be no less than 42.5% of the total oil or natural gas energy input. The minimum efficiency standard increases to 45% if the useful thermal energy is less than 15% of the total energy output of the plant.

Natural Gas Policy Act of 1978 (Public Law 95-621)

The Natural Gas Policy Act created a deregulated natural gas market for natural gas. The major objective of this regulation was to create a deregulated national market for natural gas. It provides for incremental pricing of higher cost natural gas to fluctuate with the cost of fuel oil. Cogenerators classified as qualifying facilities under PURPA are exempt from the incremental pricing schedule established for industrial customers.

Public Utility Holding Company Act of 1935 (Public Law 74-333)

The Public Utility Company Holding Act (PUHCA) of 1935 authorized the Securities and Exchange Commission (SEC) to regulate certain utility "holding companies" and their subsidiaries in a wide range of corporate transactions.

The utility industry and would-be owners of utilities lobbied Congress heavily to repeal PUHCA, claiming that it was outdated. On August 8, 2005, the Energy Policy Act of 2005 passed both houses of Congress and was signed into law, repealing PUHCA—despite consumer, environmental, union and credit rating agency objections. The repeal became effective on February 8, 2006.

SUMMARY

The dynamic process of revisions to existing codes, plus the introduction of new legislation, will continue to impact the energy industry and bring a dramatic change. Energy conservation efficiency improvement in energy products and equipment and creating new power generation supply options will be required to meet the energy demands of the twenty-first century.

A summary of historical federal energy legislation from Wickipedia below will help those looking for more detail.

* 1920—Federal Power Act
* 1946—Atomic Energy Act PL 79-585 (created the Atomic Energy Commission)
* 1954—Atomic Energy Act Amendments PL 83-703

- 1956—Colorado River Storage Project PL 84-485
- 1957—Atomic Energy Commission Acquisition of Property PL 85-162
- 1957—Price-Anderson Nuclear Industries Indemnity Act PL 85-256
- 1968—Natural Gas Pipeline Safety Act PL 90-481
- 1973—Mineral Leasing Act Amendments (Trans-Alaska Oil Pipeline Authorization) PL 93-153
- 1974—Energy Reorganization Act PL 93-438 (Split the AEC into the Energy Research and Development Administration and the Nuclear Regulatory Commission)
- 1975—Energy Policy and Conservation Act PL 94-163
- 1977—Department of Energy Organization Act PL 95-91 (Dismantled ERDA and replaced it with the Department of Energy)
- 1978—National Energy Act PL 95-617, 618, 619, 620, 621
- 1980—Energy Security Act PL 96-294
- 1989—Natural Gas Wellhead Decontrol Act PL 101-60
- 1992—Energy Policy Act of 1992 PL 102-486
- 2005—Energy Policy Act of 2005 PL 109-58
- 2007—Energy Independence and Security Act of 2007 PL 110-140
- 2008—Food, Conservation, and Energy Act of 2008 PL 110-234

Appendix A

Fundamentals: Time Value Of Money Calculations

Dr. David Pratt

Editor's Note: Appendix A provides the foundations of economic analysis and the time value of money ("A dollar today is worth more than a dollar tomorrow.") For folks who are new to financing terminol-ogy, this appendix is a good reference. Near the end of this appendix there are 38 examples about how to do economic analyses of projects!

A.1 OBJECTIVE

The objective of this appendix is to present a coherent, consistent approach to economic analysis of capital investments (energy related or other). Adherence to the concepts and methods presented will lead to sound investment decisions with respect to time value of money principles. The appendix opens with material designed to elevate the importance of life cycle cost concepts in the economic analysis of projects. The next three sections provide foundational material necessary to fully develop time value of money concepts and techniques. These sections present general characteristics of capital investments, sources of funds for capital investment, and a brief summary of tax considerations, all of which are important for economic analysis. The next two sections introduce time value of money calculations and several approaches for calculating project measures of worth based on time value of money concepts. Next the measures of worth are applied to the process of making decisions when a set of potential projects are to be evaluated. The final concept and technique section of the appendix presents material to address several special problems that may be encountered in economic analysis. This material includes, among other things, discussions of inflation, non-annual compounding of interest, and sensitivity analysis. The appendix closes with a brief summary and a list of references which can provide additional depth in many of the areas covered in the appendix.

A.2   INTRODUCTION

Capital investment decisions arise in many circumstances. The circumstances range from evaluating business opportunities to personal retirement planning. Regardless of circumstances, the basic criterion for evaluating any investment decision is that the revenues (savings) generated by the investment must be greater than the costs incurred. The number of years over which the revenues accumulate and the comparative importance of future dollars (revenues or costs) relative to present dollars are important factors in making sound investment decisions. This consideration of costs over the entire life cycle of the investment gives rise to the name "life cycle cost analysis," which is commonly used to refer to the economic analysis approach presented in this appendix. An example of the importance of life cycle costs is shown in Figure A-1, which depicts the estimated costs of owning and operating an oil-fired furnace to heat a 2,000-square-foot house in the northeast United States. Of particular note is that the initial costs represent only 23% of the total costs incurred over the life of the furnace. The life cycle cost approach provides a significantly better evaluation of long-term implications of an investment than methods which focus on first cost or near-term results.

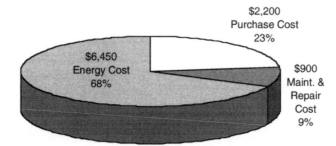

Figure A-1. 15-year life cycle costs of a heating system

Life cycle cost analysis methods can be applied to virtually any public or private business sector investment decision as well as to personal financial planning decisions. Energy related decisions, provide excellent examples for the application of this approach. Such decisions include: evaluation of alternative building designs that have different initial costs, operating and maintenance costs, and perhaps different lives; evaluation of investments to improve the thermal performance of an existing building (wall or roof insulation, window glazing); and evaluation of alternative heating, ventilating, or air conditioning systems. For federal buildings, Congress and the president have mandated, through legislation and executive order, energy conservation goals that must be met using cost-effective measures. The life cycle cost approach is mandated as the means of evaluating cost effectiveness.

A.3 GENERAL CHARACTERISTICS OF CAPITAL INVESTMENTS

A.3.1 Capital Investment Characteristics

When companies spend money, the outlay of cash can be broadly categorized into one of two classifications, expenses or capital investments. Expenses are generally those cash expenditures that are routine, on-going, and necessary for the ordinary operation of the business. Capital investments, on the other hand, are generally more strategic and have long-term effects. Decisions made regarding capital investments are usually made at higher levels within the organizational hierarchy and carry with them additional tax consequences as compared to expenses.

Three characteristics of capital investments are of concern when performing life cycle cost analysis. First, capital investments usually require a relatively large initial cost. "Relatively large" may mean several hundred dollars to a small company or many millions of dollars to a large company. The initial cost may occur as a single expenditure, such as purchasing a new heating system, or occur over a period of several years, such as designing and constructing a new building. It is not uncommon that the funds available for capital investments projects are limited. In other words, the sum of the initial costs of all the viable and attractive projects exceeds the total available funds. This creates a situation known as capital rationing that imposes special requirements on the investment analysis. This topic will be discussed in Section A.8.3.

The second important characteristic of a capital investment is that the benefits (revenues or savings) resulting from the initial cost occur in the future, normally over a period of years. The period between the initial cost and the last future cash flow is the life cycle or life of the investment. It is the fact that cash flows occur over the investment's life that requires the introduction of time value of money concepts to properly evaluate investments. If multiple investments are being evaluated and the lives of the investments are not equal, special consideration must be given to the issue of selecting an appropriate planning horizon for the analysis.

Planning horizon issues are introduced in Section A.8.5.

The last important characteristic of capital investments is that they are relatively irreversible. Frequently, after the initial investment has been made, terminating or significantly altering the nature of a capital investment has substantial (usually negative) cost consequences. This is one of the reasons that capital investment decisions are usually evaluated at higher levels of the organizational hierarchy than are operating expense decisions.

A.3.2 Capital Investment Cost Categories

In almost every case, the costs which occur over the life of a capital investment can be classified into one of the following categories:

* Initial Cost,

* Annual Expenses and Revenues,

* Periodic Replacement and Maintenance, or

* Salvage Value.

As a simplifying assumption, the cash flows which occur during a year are generally summed and regarded as a single end-of-year cash flow. While this approach does introduce some inaccuracy in the evaluation, it is generally not regarded as significantly relative to the level of estimation associated with projecting future cash flows.

Initial costs include all costs associated with preparing the investment for service. This includes purchase cost as well as installation and preparation costs. Initial costs are usually nonrecurring during the life of an investment. Annual expenses and revenues are the recurring costs and benefits generated throughout the life of the investment. Periodic replacement and maintenance costs are similar to annual expenses and revenues, except that they do not (or are not expected to) occur annually. The salvage (or residual) value of an investment is the revenue (or expense) attributed to disposing of the investment at the end of its useful life.

A.3.3 Cash Flow Diagrams

A convenient way to display the revenues (savings) and costs associated with an investment is a cash flow diagram. By using a cash flow diagram, the timing of the cash flows are more apparent, and the chances of properly applying time value of money concepts are increased. With practice, different cash flow patterns can be recognized and may suggest the most direct approach for analysis.

It is usually advantageous to determine the time frame over which the cash flows occur first. This establishes the horizontal scale of the cash flow diagram. This scale is divided into time periods that are frequently, but not always, years. Receipts and disbursements are then located on the time scale in accordance with the problem specifications. Individual outlays or receipts are indicated by drawing vertical lines appropriately placed along the time scale. The relative magnitudes can be suggested by the heights, but exact scaling generally does not enhance the meaningfulness of the diagram. Upward directed lines indicate cash inflow (revenues or savings) while downward directed lines indicate cash outflow (costs).

Figure A-2 illustrates a cash flow diagram. The cash flows depicted represent an economic evaluation of whether to choose a baseboard heating and window air conditioning system or a heat pump for a ranger's house in a national park [Fuller and Petersen, 1994]. The differential costs associated with the decision are:

- The heat pump costs (cash outflow) $1500 more than the baseboard system.

- The heat pump saves (cash inflow) $380 annually in electricity costs.

- The heat pump has a $50 higher annual maintenance cost (cash outflow).

- The heat pump has a $150 higher salvage value (cash inflow) at the end of 15 years.

- The heat pump requires $200 more in replacement maintenance (cash outflow) at the end of year 8.

Although cash flow diagrams are simply graphical representations of income and outlay, they should exhibit as much information as possible. During the analysis phase, it is useful to show the "minimum attractive rate of return" (an interest rate used to account for the time value of money within the problem) on the cash flow diagram, although this has been omitted in Figure A-2. The requirements for a good cash flow diagram are completeness, accuracy, and legibility. The measure of a successful diagram is that someone else can understand the problem fully from it.

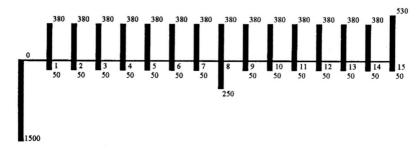

Figure A-2. Heat pump and baseboard system differential life cycle costs

A.4   SOURCES OF FUNDS

Capital investing requires a source of funds. For large companies multiple sources may be employed. The process of obtaining funds for capital investment is called financing. There are two broad sources of financial funding, debt financing and equity financing. Debt financing involves borrowing and utilizing money that is to be repaid at a later point in time. Interest is paid to the lending party for the privilege of using the money. Debt financing does not create an ownership position for the lender within the borrowing organization. The borrower is simply obligated to repay the borrowed funds, plus accrued interest according to a repayment schedule. Car loans and mortgage loans are two examples of this type of financing. The two primary sources of debt capital are loans and bonds. The cost of capital associated with debt financing is relatively easy to calculate, since interest rates and repayment schedules are usually clearly documented in the legal instruments controlling the financing arrangements. An added benefit to debt financing under current U.S. tax law (as of April 2000) is that the interest payments made by corporations on debt capital are tax deductible. This effectively lowers the cost of debt financing. For debt financing with deductible interest payments, the after-tax cost of capital is given by:

$$\text{Cost of Capital}_{\text{AFTERTAX}} =$$
$$\text{Cost of Capital}_{\text{BEFORETAX}} * (1 \text{ TaxRate}$$
where the tax rate is determined by applicable tax law.

The second broad source of funding is equity financing. Under equity financing the lender acquires an ownership (or equity) position within the borrower's organization. As a result of this ownership position, the lender has the right to participate in the financial success of the organization as a whole. The two primary sources of equity financing are stocks and retained earnings. The cost of capital associated with shares of stock is much debated within the financial community. A detailed presentation of the issues and approaches is beyond the scope of this appendix. Additional reference material can be found in Park and Sharp-Bette [1990]. One issue about which there is general agreement is that the cost of capital for stocks is higher than the cost of capital for debt financing. This is at least partially attributable to the fact that interest payments are tax deductible, while stock dividend payments are not.

If any subject is more widely debated in the financial community than the cost of capital for stocks, it is the cost of capital for retained earnings. Retained earnings are the accumulation of annual earnings surpluses that a company retains within the company's coffers rather than pays out to the stockholders as dividends. Although these earnings are held by the company, they truly belong to the stockholders. In essence the company is establishing the position that by retaining the earnings and investing them in capital projects, stockholders will achieve at least as high a return through future financial successes as they would have earned if the earnings had been paid out as dividends. Hence, one common approach to valuing the cost of capital for retained earnings is to apply the same cost of capital as for stock. This, therefore, leads to the same generally agreed result. The cost of capital for financing through retained earnings generally exceeds the cost of capital for debt financing.

In many cases the financing for a set of capital investments is obtained by packaging a combination of the above sources to achieve a desired level of available funds. When this approach is taken, the overall cost of capital is generally taken to be the weighted average cost of capital across all sources. The cost of each individual source's funds is weighted by the source's fraction of the total dollar amount available. By summing across all sources, a weighted average cost of capital is calculated, as shown in the following example:

Example 1
Determine the weighted average cost of capital for financing, which is composed of:

25% loans with a before tax cost of capital of 12%/yr and
75% retained earnings with a cost of capital of 10%/yr.
The company's effective tax rate is 34%.

Cost of Capital$_{LOANS}$ = 12% * (1 − 0.34) = 7.92%

Cost of Capital$_{RETAINEDEARNINGS}$ = 10%

Weighted Average Cost of Capital = (0.25)*7.92% + (0.75)*10.00% = 9.48%

A.5   TAX CONSIDERATIONS

A.5.1 After-tax Cash Flows
Taxes are a fact of life in both personal and business decision-making. Taxes occur in many forms and are primarily designed to generate revenues for governmental entities ranging from local authorities to the federal government. A few of the most common forms of taxes are income taxes, ad valorem taxes, sales taxes, and excise taxes. Cash flows used for economic analysis should always be adjusted for the combined impact of all relevant taxes. To do otherwise ignores the significant impact that taxes have on economic decision-making. Tax laws and regulations are complex and intricate. A detailed treatment of tax considerations as they apply to economic analysis is beyond the scope of this appendix and generally requires the assistance of a professional with specialized training in the subject. A high level summary of concepts and techniques that concentrate on federal income taxes is presented in the material which follows. The focus is on federal income taxes, since they impact most decisions and have relatively wide and general application.

The amount of federal taxes due are determined based on a tax rate multiplied by a taxable income. The rates (as of April 2000) are determined based on tables of rates published under the Omnibus Reconciliation Act of 1993 as shown in Table A-1. Depending on income range, the marginal tax rates vary from 15% of taxable income to 39% of taxable income. Taxable income is calculated by subtracting allowable deductions from gross income. Gross income is generated when a company sells its prod-

Table A-1. Federal tax rates based on the Omnibus Reconciliation Act of 1993

Taxable Income (TI)	Taxes Due	Marginal Tax Rate
$0 < TI ≤ $50,000	0.15*TI	0.15
$50,000 < TI ≤ $75,000	$7,500+0.25(TI-$50,000)	0.25
$75,000 < TI ≤ $100,000	$13,750+0.34(TI-$75,000)	0.34
$100,000 < TI ≤ $335,000	$22,250+0.39(TI-$100,000)	0.39
$335,000 < TI ≤ $10,000,000	$113,900+0.34(TI-$335,000)	0.34
$10,000,000 < TI ≤ $15,000,000	$3,400,000+0.35(TI-$10,000,000)	0.35
$15,000,000 < TI ≤ $18,333,333	$5,150,000+0.38(TI-$15,000,000)	0.38
$18,333,333 < TI	$6,416,667+0.35(TI-$18,333,333)	0.35

uct or service. Allowable deductions include salaries and wages, materials, interest payments, and depreciation, as well as other costs of doing business as detailed in the tax regulations.

The calculation of taxes owed and after-tax cash flows (ATCF) requires knowledge of:

- Before Tax Cash Flows (BTCF), the net project cash flows before the consideration of taxes due, loan payments, and bond payments;

- Total loan payments attributable to the project, including a breakdown of principal and interest components of the payments;

- Total bond payments attributable to the project, including a breakdown of the redemption and interest components of the payments; and

- Depreciation allowances attributable to the project.

Given the availability of the above information, the procedure to determine the ATCF on a year-by-year basis proceeds by using the following calculation for each year:

- Taxable Income = BTCF – Loan Interest – Bond Interest – Depreca-
tion

- Taxes = Taxable Income * Tax Rate

- ATCF = BTCF – Total Loan Payments – Total Bond Payments – Taxes

An important observation is that depreciation reduces taxable income (hence, taxes) but does not directly enter into the calculation of ATCF since it is not a true cash flow. It is not a true cash flow because no cash changes hands. Depreciation is an accounting concept designed to stimulate business by reducing taxes over the life of an asset. The next section provides additional information about depreciation.

A.5.2 Depreciation

Most assets used in the course of a business decrease in value over time. U.S. federal income tax law permits reasonable deductions from taxable income to allow for this. These deductions are called depreciation allowances. To be depreciable, an asset must meet three primary conditions: (1) it must be held by the business for the purpose of producing income, (2) it must wear out or be consumed in the course of its use, and (3) it must have a life longer than a year.

Many methods of depreciation have been allowed under U.S. tax law over the years. Among these methods are straight line, sum-of-the-years digits, declining balance, and the accelerated cost recovery system. Descriptions of these methods can be found in many references, including economic analysis text books [White, et al., 1998]. The method currently used for depreciation of assets placed in service after 1986 is the Modified Accelerated Cost Recovery System (MACRS). Determination of the allowable MACRS depreciation deduction for an asset is a function of (1) the asset's property class, (2) the asset's basis, and (3) the year within the asset's recovery period for which the deduction is calculated.

Eight property classes are defined for assets which are depreciable under MACRS. The property classes and several examples of property that fall into each class are shown in Table A-2. Professional tax guidance is recommended to determine the MACRS property class for a specific asset.

The basis of an asset is the cost of placing the asset in service. In most cases, the basis includes the purchase cost of the asset plus the costs

Table A-2. MACRS property classes

Property Class	Example Assets
3-Year Property	special handling devices for food special tools for motor vehicle manufacturing
5-Year Property	computers and office machines general purpose trucks
7-Year Property	office furniture most manufacturing machine tools
10-Year Property	tugs & water transport equipment petroleum refining assets
15-Year Property	fencing and landscaping cement manufacturing assets
20-Year Property	farm buildings utility transmission lines and poles
27.5-Year Residential Rental Property	rental houses and apartments
31.5-Year Nonresidential Real Property	business buildings

necessary to place the asset in service (e.g., installation charges).

Given an asset's property class and its depreciable basis, the depreciation allowance for each year of the asset's life can be determined from tabled values of MACRS percentages. The MACRS percentages specify the percentage of an asset's basis that are allowable as deductions during each year of an asset's recovery period. The MACRS percentages by recovery year (age of the asset) and property class are shown in Table A-3.

Table A-3. MACRS percentages by recovery year and property class

Recovery Year	3-Year Property	5-Year Property	7-Year Property	10-Year Property	15-Year Property	20-Year Property
1	33.33%	20.00%	14.29%	10.00%	5.00%	3.750%
2	44.45%	32.00%	24.49%	18.00%	9.50%	7.219%
3	14.81%	19.20%	17.49%	14.40%	8.55%	6.677%
4	7.41%	11.52%	12.49%	11.52%	7.70%	6.177%
5		11.52%	8.93%	9.22%	6.93%	5.713%
6		5.76%	8.92%	7.37%	6.23%	5.285%
7			8.93%	6.55%	5.90%	4.888%
8			4.46%	6.55%	5.90%	4.522%
9				6.56%	5.91%	4.462%
10				6.55%	5.90%	4.461%
11				3.28%	5.91%	4.462%
12					5.90%	4.461%
13					5.91%	4.462%
14					5.90%	4.461%
15					5.91%	4.462%
16					2.95%	4.461%
17						4.462%
18						4.461%
19						4.462%
20						4.461%
21						2.231%

Example 2

Determine depreciation allowances during each recovery year for a MACRS 5-year property with a basis of $10,000.

Year 1 deduction: $10,000 * 20.00% = $2,000
Year 2 deduction: $10,000 * 32.00% = $3,200
Year 3 deduction: $10,000 * 19.20% = $1,920
Year 4 deduction: $10,000 * 11.52% = $1,152
Year 5 deduction: $10,000 * 11.52% = $1,152
Year 6 deduction: $10,000 * 5.76% = $576

The sum of the deductions calculated in Example 2 is $10,000, which means that the asset is "fully depreciated" after six years. Though not shown here, tables similar to Table A-3 are available for the 27.5-year and 31.5-year property classes. Their usage is similar to that outlined above, except that depreciation is calculated monthly rather than annually.

A.6   TIME VALUE OF MONEY CONCEPTS

A.6.1 Introduction

Most people have an intuitive sense of the time value of money. Given a choice between $100 today and $100 one year from today, almost everyone would prefer the $100 today. Why is this the case? Two primary factors lead to this time preference associated with money; interest and inflation. Interest is the ability to earn a return on money which is loaned rather than consumed. By taking the $100 today and placing it in an interest bearing bank account (i.e., loaning it to the bank), one year from today an amount greater than $100 would be available for withdrawal. Thus, taking the $100 today and loaning it to earn interest, generates a sum greater than $100 one year from today and is thus preferred. The amount in excess of $100 that would be available depends upon the interest rate being paid by the bank. The next section develops the mathematics of the relationship between interest rates and the timing of cash flows.

The second factor which leads to the time preference associated with money is inflation. Inflation is a complex subject but in general can be described as a decrease in the purchasing power of money. The impact of inflation is that the "basket of goods" a consumer can buy today with $100 contains more than the "basket" the consumer could buy one year from today. This decrease in purchasing power is the result of inflation. The subject of inflation is addressed in Section A.9.4.

A.6.2 The Mathematics of Interest

The mathematics of interest must account for the amount and timing of cash flows. The basic formula for studying and understanding interest calculations is:

$$F_n = P + I_n$$

where: F_n = a future amount of money at the *end* of the nth year

   P = a present amount of money at the beginning of the year which is n years prior to F_n

   I_n = the amount of accumulated interest over n years

   n = the number of years between P and F

The goal of studying the mathematics of interest is to develop a formula for F_n that is expressed only in terms of the present amount P, the annual interest rate i, and the number of years n. There are two major approaches for determining the value of I_n: simple interest and compound interest. Under simple interest, interest is earned (charged) only on the original amount loaned (borrowed). Under compound interest, interest is earned (charged) on the original amount loaned (borrowed) plus any interest accumulated from previous periods.

A.6.3 Simple Interest

For simple interest, interest is earned (charged) only on the original principal amount at the rate of i% per year (expressed as i%/yr). Table A-4 illustrates the annual calculation of simple interest. In Table A-4 and the formulas which follow, the interest rate i is to be expressed as a decimal amount (e.g., 8% interest is expressed as 0.08).

At the beginning of year 1 (end of year 0), P dollars (e.g., $100) are deposited in an account earning i%/yr (e.g., 8%/yr or 0.08) simple interest. Under simple compounding, during year 1 the P dollars ($100) earn

Table A-4. The mathematics of simple interest

Year (t)	Amount At Beginning Of Year	Interest Earned During Year	Amount At End Of Year (F_t)
0	-	-	P
1	P	Pi	P + Pi = P (1 + i)
2	P (1 + i)	Pi	P (1+ i) + Pi = P (1 + 2i)
3	P (1 + 2i)	Pi	P (1+ 2i) + Pi = P (1 + 3i)
n	P (1 + (n-1)i)	Pi	P (1+ (n-1)i) + Pi = P (1 + ni)

P*i dollars ($100*0.08 = $8) of interest. At the end of the year 1 the balance in the account is obtained by adding P dollars (the original principal, $100) plus P*i (the interest earned during year 1, $8) to obtain P+P*i ($100+$8 = $108). Through algebraic manipulation, the end of year 1 balance can be expressed mathematically as P*(1+i) dollars ($100*1.08 = $108).

The beginning of year 2 is the same point in time as the end of year 1 so the balance in the account is P*(1+i) dollars ($108). During year 2 the account again earns P*i dollars ($8) of interest, since under simple compounding interest is paid only on the original principal amount P ($100). Thus at the end of year 2, the balance in the account is obtained by adding P dollars (the original principal) plus P*i (the interest from year 1) plus P*i (the interest from year 2) to obtain P+P*i+P*i ($100+$8+$8 = $116). After some algebraic manipulation, this can conveniently be written mathematically as P*(1+2*i) dollars ($100*1.16 = $116).

Table A-4 extends the above logic to year 3 and then generalizes the approach for year n. If we return our attention to our original goal of developing a formula for F_n that is expressed only in terms of the present amount P, the annual interest rate i, and the number of years n, the above development and Table A-4 results can be summarized as follows:

For Simple Interest
$$F_n = P \, (1+n*i)$$

Example 3

Determine the balance which will accumulate at the end of year 4 in an account which pays 10%/yr simple interest if a deposit of $500 is made today.

$F_n = P * (1 + n*i)$

$F_4 = 500 * (1 + 4*0.10)$

$F_4 = 500 * (1 + 0.40)$

$F_4 = 500 * (1.40)$

$F_4 = \$700$

A.6.4 Compound Interest

For compound interest, interest is earned (charged) on the original principal amount, plus any accumulated interest from previous years at the rate of i% per year (i%/yr). Table A-5 illustrates the annual calculation of compound interest. In the Table A-5 and the formulas which follow, i is expressed as a decimal amount (i.e., 8% interest is expressed as 0.08).

At the beginning of year 1 (end of year 0), P dollars (e.g., $100) are deposited in an account earning i%/yr (e.g., 8%/yr or 0.08) compound interest. Under compound interest, during year 1 the P dollars ($100) earn P*i dollars ($100*0.08 = $8) of interest. Notice that this the same as the amount earned under simple compounding. This result is expected since the interest earned in previous years is zero for year 1. At the end of the year 1 the balance in the account is obtain by adding P dollars (the original principal, $100) plus P*i (the interest earned during year 1, $8) to obtain P+P*i ($100+$8 = $108). Through algebraic manipulation, the end of year 1 balance can be expressed mathematically as P*(1+i) dollars ($100*1.08 = $108).

During year 2 and subsequent years, we begin to see the power (if you are a lender) or penalty (if you are a borrower) of compound interest over simple interest. The beginning of year 2 is the same point in time as the end of year 1 so the balance in the account is P*(1+i) dollars ($108). During year 2 the account earns i% interest on the original principal, P dollars ($100), *and* it earns i% interest on the accumulated interest from year 1, P*i dollars ($8). Thus the interest earned in year 2 is [P+P*i]*i dol-

Table A-5. The Mathematics of Compound Interest

Year (t)	Amount At Beginning Of Year	Interest Earned During Year	Amount At End Of Year (F_t)
0	-	-	P
1	P	Pi	$P + Pi$ $= P(1+i)$
2	$P(1+i)$	$P(1+i)i$	$P(1+i) + P(1+i)i$ $= P(1+i)(1+i)$ $= P(1+i)^2$
3	$P(1+i)^2$	$P(1+i)^2 i$	$P(1+i)^2 + P(1+i)^2 i$ $= P(1+i)^2(1+i)$ $= P(1+i)^3$
n	$P(1+i)^{n-1}$	$P(1+i)^{n-1} i$	$P(1+i)^{n-1} + P(1+i)^{n-1} i$ $= P(1+i)^{n-1}(1+i)$ $= P(1+i)^n$

lars ([$100+$8]*0.08 = $8.64). The balance at the end of year 2 is obtained by adding P dollars (the original principal) plus P*i (the interest from year 1) plus [P+P*i]*i (the interest from year 2) to obtain P+P*i+[P+P*i]*i dollars ($100+$8+$8.64 = $116.64). After some algebraic manipulation, this can be conveniently written mathematically as $P*(1+i)^n$ dollars ($100*1.082 = $116.64).

Table A-5 extends the above logic to year 3 and then generalizes the approach for year n. If we return our attention to our original goal of developing a formula for F_n that is expressed only in terms of the present amount P, the annual interest rate i, and the number of years n, the above development and Table A-5 results can be summarized as follows:

For Compound Interest
$$F_n = P\,(1+i)^n$$

Example 4

Repeat Example 3 using compound interest rather than simple interest.

$F_n = P * (1 + i)^n$

$F_4 = 500 * (1 + 0.10)^4$

$F_4 = 500 * (1.10)^4$

$F_4 = 500 * (1.4641)$

$F_4 = \$732.05$

Notice that the balance available for withdrawal is higher under compound interest ($732.05 > $700.00). This is due to earning interest on principal plus interest rather than earning interest on just original principal. Since compound interest is by far more common in practice than simple interest, the remainder of this appendix is based on *compound interest* unless explicitly stated otherwise.

A.6.5 Single Sum Cash Flows

Time value of money problems involving compound interest are common. Because of this frequent need, tables of compound interest time value of money factors can be found in most books and reference manuals that deal with economic analysis. The factor $(1+i)^n$ is known as the *single sum, future worth factor*, or the *single payment, compound amount factor*. This factor is denoted (F | P,i,n), where F denotes a future amount, P denotes a present amount, i is an interest rate (expressed as a percentage amount), and n denotes a number of years. The factor $(F \mid P,i,n)$ is read "to find F given P at i% for n years." Tables of values of $(F \mid P,i,n)$ for selected values of i and n are provided in Appendix 4A. The tables of values in Appendix 4A are organized such that the annual interest rate (i) determines the appropriate page, the time value of money factor $(F \mid P)$ determines the appropriate column, and the number of years (n) determines the appropriate row.

Example 5
   Repeat Example 4 using the single sum, future worth factor.
$F_n = P * (1 + i)^n$
$F_n = P * (F \mid P,i,n)$
$F_4 = 500 * (F \mid P,10\%,4)$
$F_4 = 500 * (1.4641)$
$F_4 = 732.05$

   The above formulas for compound interest allow us to solve for an unknown F, given P, i, and n. What if we want to determine P with known values of F, i, and n? We can derive this relationship from the compound interest formula above:

$$F_n - P (1 + i)^n$$
Dividing both sides by $(1+i)^n$ yields

$$P = \frac{F_n}{(1 + i)^n}$$

which can be rewritten as
$$P = F_n (1+i)^{-n}$$

   The factor $(1+i)^{-n}$ is known as the single sum; present worth factor; or the single payment, present worth factor. This factor is denoted $(P \mid F,i,n)$ and is read "to find P given F at i% for n years." Tables of $(P \mid F,i,n)$ are provided in Appendix 4A.

Example 6
   To accumulate $1000 five years from today in an account earning 8%/yr compound interest, how much must be deposited today?

$$P = F_n * (1 + i)^{-n}$$
$$P = F5 * (P \mid F,i,n)$$
$$P = 1000 * (P \mid F,8\%,5)$$
$$P = 1000 * (0.6806)$$
$$P = 680.60$$

To verify your solution, try multiplying 680.60 * (F I P,8%,5). What would expect for a result? (Answer: $1000) If your still not convinced, try building a table like Table A-5 to calculate the year end balances each year for five years.

A.6.6 Series Cash Flows

Having considered the transformation of a single sum to a future worth when given a present amount and vice versa, let us generalize to a series of cash flows. The future worth of a series of cash flows is simply the sum of the future worths of each individual cash flow. Similarly, the present worth of a series of cash flows is the sum of the present worths of the individual cash flows.

Example 7

Determine the future worth (accumulated total) at the end of seven years in an account that earns 5%/yr if a $600 deposit is made today and a $1000 deposit is made at the end of year two?

for the $600 deposit, n = 7 (years between today and end of year 7)

for the $1000 deposit, n = 5 (years between end of year 2 and end of year 7)

$$F7 = 600 * (F \mid P,5\%,7) + 1000 * (F \mid P,5\%,5)$$
$$F7 = 600 * (1.4071) + 1000 * (1.2763)$$
$$F7 = 844.26 + 1276.30 = \$2120.56$$

Example 8

Determine the amount that would have to be deposited today (present worth) in an account paying 6%/yr interest if you want to withdraw $500 four years from today and $600 eight years from today (leaving zero in the account after the $600 withdrawal).

For the $500 deposit n = 4, for the $600 deposit n = 8

P = 500 * (P | F,6%,4) + 600 * (P | F,6%,8)

P = 500 * (0.7921) + 600 * (0.6274)

P = 396.05 + 376.44 = $772.49

A.6.7 Uniform Series Cash Flows

A uniform series of cash flows exists when the cash flows are in a series, occur every year, and are all equal in value. Figure A-3 shows the cash flow diagram of a uniform series of withdrawals. The uniform series has length 4 and amount 2000. If we want to determine the amount of money that would have to be deposited today to support this series of withdrawals, starting one year from today, we could use the approach illustrated in Example 8 above to determine a present worth component for each individual cash flow. This approach would require us to sum the following series of factors (assuming the interest rate is 9%/yr):

P = 2000*(P | F,9%,1) + 2000*(P | F,9%,2) +
     2000*(P | F,9%,3) + 2000*(P | F,9%,4)

After some algebraic manipulation, this expression can be restated as:

P = 2000*[(P | F,9%,1) + (P | F,9%,2) +
     (P | F,9%,3) + (P | F,9%,4)]

P = 2000*[(0.9174) + (0.8417) + (0.7722) + (0.7084)]

P = 2000*[3.2397] = $6479.40

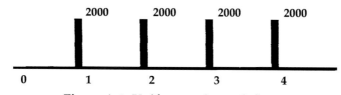

Figure A-3. Uniform series cash flow

Fortunately, uniform series occur frequently enough in practice to justify tabulating values to eliminate the need to repeatedly sum a series of (P | F,i,n) factors. To accommodate uniform series factors, we need to add a new symbol to our time value of money terminology in addition to the single sum symbols P and F. The symbol "A" is used to designate a uniform series of cash flows. When dealing with uniform series cash flows, the symbol A represents the amount of each annual cash flow and n represents the number of cash flows in the series. The factor (P | A,i,n) is known as the *uniform series, present worth factor* and is read "to find P given A at i% for n years." Tables of (P | A,i,n) are provided in Appendix 4A. An algebraic expression can also be derived for the (P | A,i,n) factor which expresses P in terms of A, i, and n. The derivation of this formula is omitted here, but the resulting expression is shown in the summary table (Table A-6) at the end of this section.

An important observation when using a (P | A,i,n) factor is that the "P" resulting from the calculation occurs one period prior to the first "A" cash flow. In our example the first withdrawal (the first "A") occurred one year after the deposit ("P"). Restating the example problem above using a (P | A,i,n) factor, it becomes:

$P = A * (P | A,i,n)$

$P = 2000 * (P | A,9\%,4)$

$P = 2000 * (3.2397) = \$6479.40$

This result is identical (as expected) to the result using the (P | F,i,n) factors. In both cases the interpretation of the result is: If we deposit $6479.40 in an account paying 9%/yr interest, we could make withdrawals of $2000 per year for four years, starting one year after the initial deposit, to deplete the account at the end of 4 years.

The reciprocal relationship between P and A is symbolized by the factor (A | P,i,n) and is called the *uniform series, capital recovery factor*. Tables of (A | P,i,n) are provided in Appendix 4A, and the algebraic expression for (A | P,i,n) is shown in Table A-6 at the end of this section. This factor enables us to determine the amount of the equal annual withdrawals "A" (starting one year after the deposit) that can be made from an initial deposit of "P."

Example 9

Determine the equal annual withdrawals that can be made for 8 years from an initial deposit of $9000 in an account that pays 12%/yr. The first withdrawal is to be made one year after the initial deposit.

$A = P * (A | P,12\%,8)$

$A = 9000 * (0.2013)$

A = $1811.70

Factors are also available for the relationships between a future worth (accumulated amount) and a uniform series. The factor (F | A,i,n) is known as the *uniform series future worth* factor and is read "to find F given A at i% for n years." The reciprocal factor, (A | F,i,n), is known as the *uniform series sinking fund* factor and is read "to find A given F at i% for n years." An important observation when using an (F | A,i,n) factor or an (A | F,i,n) factor is that the "F" resulting from the calculation occurs at the same point in time as to the last "A" cash flow. The algebraic expressions for (A | F,i,n) and (F | A,i,n) are shown in Table 6 at the end of this section.

Example 10

If you deposit $2000 per year into an individual retirement account starting on your 24th birthday, how much will have accumulated in the account at the time of your deposit on your 65th birthday? The account pays 6% / yr.

n = 42 (birthdays between 24th and 65th, inclusive)

F = A * (F | A,6%,42)

F = 2000 * (175.9505) = $351,901

Example 11

If you want to be a millionaire on your 65th birthday, what equal annual deposits must be made in an account starting on your 24th birthday? The account pays 10% / yr.

n = 42 (birthdays between 24th and 65th, inclusive)

A = F * (A | F,10%,42)

A = 1000000 * (0.001860) = $1860

A.6.8 Gradient Series

A gradient series of cash flows occurs when the value of a given cash flow is greater than the value of the previous period's cash flow by a constant amount. The symbol used to represent the constant increment is G. The factor (P | G,i,n) is known as the *gradient series, present worth factor*.

Tables of (P | G,i,n) are provided in Appendix 4A. An algebraic expression can also be derived for the (P | G,i,n) factor, which expresses P in terms of G, i, and n. The derivation of this formula is omitted here, but the resulting expression is shown in the summary table (Table A-6) at the end of this section.

It is not uncommon to encounter a cash flow series that is the sum of a uniform series and a gradient series. Figure A-4 illustrates such a series. The uniform component of this series has a value of 1000 and the gradient series has a value of 500. By convention the first element of a gradient series has a zero value. Therefore, in Figure A-4, both the uniform series and the gradient series have length four (n = 4). Like the uniform series factor, the "P" calculated by a (P | G,i,n) factor is located one period before the first element of the series (which is the zero element for a gradient series).

Example 12

Assume you wish to make the series of withdrawals illustrated in Figure A-4 from an account which pays 15% / yr. How much money would you have to deposit today such that the account is depleted at the time of the last withdrawal?

This problem is best solved by recognizing that the cash flows are a combination of a uniform series of value 1000 and length 4 (starting at time = 1), plus a gradient series of size 500 and length 4 (starting at time = 1).

P = A * (P | A,15%,4) + G * (P | G,15%,4)
P = 1000 * (2.8550) + 500 * (3.7864)
P = 2855.00 + 1893.20 = $4748.20

Occasionally it is useful to convert a gradient series to an equivalent

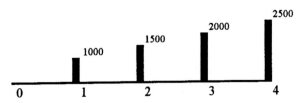

Figure A-4. Combined uniform series and gradient series cash flow

uniform series of the same length. Equivalence in this context means that the present value (P) calculated from the gradient series is numerically equal to the present value (P) calculated from the uniform series. One way to accomplish this task with the time value of money factors we have already considered is to convert the gradient series to a present value using a (P|G,i,n) factor and then convert this present value to a uniform series using an (A|P,i,n) factor. In other words:

$$A = [G * (P|G,i,n)] * (A|P,i,n)$$

An alternative approach is to use a factor known as the *gradient-to-uniform series conversion factor*, symbolized by (A|G,i,n). Tables of (A|G,i,n) are provided in Appendix 4A. An algebraic expression can also be derived for the (A|G,i,n) factor, which expresses A in terms of G, i, and n. The derivation of this formula is omitted here, but the resulting expression is shown in the summary table (Table A-6) at the end of this section.

A.6.9 Summary of Time Value of Money Factors

Table A-6 summarizes the time value of money factors introduced in this section. Time value of money factors are useful in economic analysis, because they provide a mechanism to accomplish two primary func tions· (1) they allow us to replace a cash flow at one point in time with an equivalent cash flow (in a time value of money sense) at a different point in time and (2) they allow us to convert one cash flow pattern to another (e.g., convert a single sum of money to an equivalent cash flow series or convert a cash flow series to an equivalent single sum). The usefulness of these two functions when performing economic analysis of alternatives will become apparent in Sections A.7 and A.8, which follow.

A.6.10 The Concepts of Equivalence and Indifference

Up to this point the term "equivalence" has been used several times but never fully defined. It is appropriate at this point to formally define equivalence, as well as a related term, "indifference."

In economic analysis, "equivalence" means "the state of being equal in value." The concept is primarily applied to the comparison of two or more cash flow profiles. Specifically, two (or more) cash flow profiles are equivalent if their time value of money worths at a common point in time are equal.

Table A-6 Summary of discrete compounding time value of money factors

To Find	Given	Factor	Symbol	Name
P	F	$(1+i)^{-n}$	$(P\,|\,F,i,n)$	Single Payment, Present Worth Factor
F	P	$(1+i)^{n}$	$(F\,|\,P,i,n)$	Single Payment, Compound Amount Factor
P	A	$\dfrac{(1+i)^{n}-1}{i(1+i)^{n}}$	$(P\,|\,A,i,n)$	Uniform Series, Present Worth Factor
A	P	$\dfrac{i(1+i)^{n}}{(1+i)^{n}-1}$	$(A\,|\,P,i,n)$	Uniform Series, Capital Recovery Factor
F	A	$\dfrac{(1+i)^{n}-1}{i}$	$(F\,|\,A,i,n)$	Uniform Series, Compound Amount Factor
A	F	$\dfrac{i}{(1+i)^{n}-1}$	$(A\,|\,F,i,n)$	Uniform Series, Sinking Fund Factor
P	G	$\dfrac{1-(1+ni)(1+i)^{-n}}{i^{2}}$	$(P\,|\,G,i,n)$	Gradient Series, Present Worth Factor
A	G	$\dfrac{(1+i)^{n}-(1+ni)}{i\left[(1+i)^{n}-1\right]}$	$(A\,|\,G,i,n)$	Gradient Series, Uniform Series Factor

Question: Are the following two cash flows equivalent at 15%/yr?
Cash Flow 1: Receive $1,322.50 two years from today
Cash Flow 2: Receive $1,000.00 today

Analysis Approach 1: Compare worths at t = 0 (present worth).
PW(1) = 1,322.50*(P | F,15,2) = 1322.50*0.756147 = 1,000 PW(2) = 1,000
Answer: Cash Flow 1 and Cash Flow 2 are equivalent.

Analysis Approach 2: Compare worths at t = 2 (future worth).
FW(1) = 1,322.50
FW(2) = 1,000*(F | P,15,2) = 1,000*1.3225 = 1,322.50
Answer: Cash Flow 1 and Cash Flow 2 are equivalent.

Generally the comparison (hence the determination of equivalence) for the two cash flow series in this example would be made as present worths (t = 0) or future worths (t = 2), but the equivalence definition holds regardless of the point in time chosen. For example:

Analysis Approach 3: Compare worths at t = 1.
W1(1) = 1,322.50*(P | F,15,1)
        = 1,322.50*0.869565 = 1,150.00
W1(2) = 1,000*(F | P,15,1) = 1,000*1.15 = 1,150.00
Answer: Cash Flow 1 and Cash Flow 2 are equivalent.

Thus, the selection of the point in time, t, at which to make the comparison is completely arbitrary. Clearly, however, some choices are more intuitively appealing than others (t = 0 and t = 2 in the above example).

In economic analysis, "indifference" means "to have no preference." The concept is primarily applied in the comparison of two or more cash flow profiles. Specifically, a potential investor is indifferent between two (or more) cash flow profiles if they are equivalent.

Question: Given the following two cash flows at 15%/yr, which do you
    prefer?
Cash Flow 1: Receive $1,322.50 two years from today
Cash Flow 2: Receive $1,000.00 today
Answer: Based on the equivalence calculations above, given these two
    choices, an investor is indifferent.

The concept of equivalence can be used to break a large, complex problem into a series of smaller, more manageable ones. This is done by taking advantage of the fact that, in calculating the economic worth of a cash flow profile, any part of the profile can be replaced by an equivalent representation at an arbitrary point in time without altering the worth of the profile.

Question: You are given a choice between (1) receiving P dollars today or (2) receiving the cash flow series illustrated in Figure A-5. What must the value of P be for you to be indifferent between the two choices if i = 12%/yr?

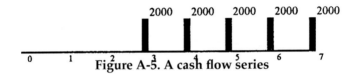

Figure A-5. A cash flow series

Analysis Approach: To be indifferent between the choices, P must have a value such that the two alternatives are equivalent at 12%/yr. If we select t = 0 as the common point in time upon which to base the analysis (present worth approach), then the analysis proceeds as follows:

PW(Alt 1) = P

Because P is already at t = 0 (today), no time value of money factors are involved.

PW(Alt 2)

Step 1— Replace the uniform series (t = 3 to 7) with an equivalent single sum, V_2, at t = 2 (one period before the first element of the series). V_2 = 2,000 * (P | A,12%,5) = 2,000 * 3.6048 = 7,209.60

Step 2— Replace the single sum V_2, with an equivalent value V_0 at t = 0. PW(Alt 2) = V_0 = V_2 * (P | F,12,2) = 7,209.60 * 0.7972 = 5,747.49

Answer: To be indifferent between the two alternatives, they must be equivalent at t = 0. To be equivalent, P must have a value of $5,747.49.

A.7 PROJECT MEASURES OF WORTH

A.7.1 Introduction

In this section measures of worth for investment projects are introduced. The measures are used to evaluate the attractiveness of a single investment opportunity. The measures to be presented are (1) present worth, (2) annual worth, (3) internal rate of return, (4) savings investment ratio, and (5) payback period. All but one of these measures of worth require an interest rate to calculate the worth of an investment. This interest rate is commonly referred to as the minimum attractive rate of return (MARR). There are many ways to determine a value of MARR for investment analysis, and no one way is proper for all applications. One principle is, however, generally accepted. MARR should always exceed the cost of capital as described in Section A.4, Sources of Funds, presented earlier in this appendix.

In all of the measures of worth below, the following conventions are used for defining cash flows. At any given point in time ($t = 0, 1, 2,..., n$), there may exist both revenue (positive) cash flows, R_t, and cost (negative) cash flows, C_t. The net cash flow at t, A_t, is defined as $R_t - C_t$.

A.7.2 Present Worth

Consider again the cash flow series illustrated in Figure A-5. If you were given the opportunity to "buy" that cash flow series for $5,747.49, would you be interested in purchasing it? If you expected to earn a 12% / yr return on your money (MARR = 12%), based on the analysis in the previous section, your conclusion should be that you are indifferent between (1) retaining your $5,747.49 and (2) giving up your $5,747.49 in favor of the cash flow series. Figure A-6 illustrates the net cash flows of this second investment opportunity.

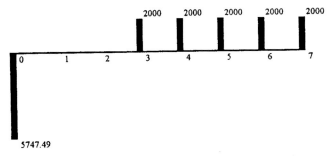

Figure A-6 An investment opportunity

What value would you expect if we calculated the present worth (equivalent value of all cash flows at t = 0) of Figure A-6? We must be careful with the signs (directions) of the cash flows in this analysis since some represent cash outflows (downward) and some represent cash inflows (upward).

$$PW = -5747.49 + 2000*(P \mid A,12\%,5)*(P \mid F,12\%,2)$$
$$PW = -5747.49 + 2000*(3.6048)*(0.7972)$$
$$PW = -5747.49 + 5747.49 = \$0.00$$

The value of zero for present worth indicates indifference regarding this investment opportunity. We would just as soon do nothing (i.e., retain our $5747.49) as invest in the opportunity.

What if the same returns (future cash inflows) were offered for a $5000 investment (t = 0 outflow)? Would this be more, or less attractive? Hopefully, after a little reflection, it is apparent that this would be a more attractive investment, because you are getting the same returns but paying less than the indifference amount for them. What happens if we calculate the present worth of this new opportunity?

$$PW = -5000 + 2000*(P \mid A,12\%,5)*(P \mid F,12\%,2)$$
$$PW = -5000 + 2000*(3.6048)*(0.7972)$$
$$PW = -5000.00 + 5747.49 = \$747.49$$

The positive value of present worth indicates an attractive investment. If we repeat the process with an initial cost greater than $5747.49, it should come as no surprise that the present worth will be negative, indicating an unattractive investment.

The concept of present worth as a measure of investment worth can be generalized as follows:

<u>Measure of Worth</u>: Present Worth

<u>Description</u>: All cash flows are converted to a single sum equivalent at time zero using i = MARR.

<u>Calculation Approach</u>: $PW = \sum_{t=1}^{n} A_t(P \mid F,i,t)$

<u>Decision Rule</u>: If PW ≥0, then the investment is attractive.

Example 13

Installing thermal windows on a small office building is estimated to cost $10,000. The windows are expected to last six years and have no salvage value at that time. The energy savings from the windows are expected to be $2525 each year for the first three years, and $3840 for each of the remaining three years. If MARR is 15%/yr and the present worth measure of worth is to be used, is this an attractive investment?

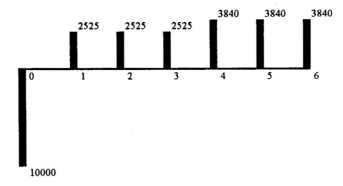

Figure A-7. Thermal windows investment

The cash flow diagram for the thermal windows is shown in Figure A-7.

PW =

    −10000+2525*(P | F,15%,1)+2525*(P | F,15%,2)
    +2525*(P | F,15%,3)+3840*(P | F,15%,4)+
    3840*(P | F,15%,5)+
    3840*(P | F,15%,6)

PW =

    −10000+2525*(0.8696)+2525*(0.7561)
    +2525*(0.6575)+
    3840*(0.5718)+3840*(0.4972)+
    3840*(0.4323)

PW =

    −10000+2195.74+1909.15+1660.19+2195.71
    +1909.25+1660.03

PW = $1530.07

<u>Decision</u>: PW≥0 ($1530.07≥0.0); therefore, the window investment is attractive.

An alternative (and simpler) approach to calculating PW is obtained by recognizing that the savings cash flows are two uniform series, one of value $2525 and length 3 starting at t = 1, and one of value $3840 and length 3 starting at t = 4.

PW =   −10000+2525*(P | A,15%,3)+3840*
        (P | A,15%,3)*(P | F,15%,3)

PW =   −10000+2525*(2.2832)+3840*(2.2832)*
        (0.6575) = $1529.70

<u>Decision</u>: PW≥0 ($1529.70>0.0); therefore, the window investment is attractive.

The slight difference in the PW values is caused by the accumulation of round off errors as the various factors are rounded to four places to the right of the decimal point.

A.7.3 Annual Worth

An alternative to present worth is annual worth. The annual worth measure converts all cash flows to an equivalent uniform annual series of cash flows over the investment life, using i = MARR. The annual worth measure is generally calculated by first calculating the present worth measure and then multiplying by the appropriate (A | P,i,n) factor. A thorough review of the tables in Appendix 4A or the equations in Table A-6 leads to the conclusion that for all values of i (i>0) and n (n>0), the value of (A | P,i,n) is greater than zero. Hence,

if PW>0, then AW>0;

if PW<0, then AW<0; and

if PW = 0, then AW = 0,

because the only difference between PW and AW is multiplication by a positive, non-zero value, namely (A | P,i,n). The decision rule for investment attractiveness for PW and AW are identical: positive values indicate an attractive investment; negative values indicate an unattractive investment; zero indicates indifference. Frequently the only reason for choosing between AW and PW as a measure of worth in an analysis is the preference of the decision maker.

The concept of annual worth as a measure of investment worth can be generalized as follows:

Measure of Worth: Annual Worth

Description: All cash flows are converted to an equivalent uniform annual series of cash flows over the planning horizon, using i = MARR.

Calculation Approach: AW = PW (A | P,i,n)

Decision Rule: If AW ≥0, then the investment is attractive.

Example 14

Reconsider the thermal window data of Example 13. If the annual worth measure of worth is to be used, is this an attractive investment?

AW = PW (A | P,15%,6)
AW = 1529.70 (0.2642) = $404.15/yr
Decision: AW ≥0 ($404.15>0.0); therefore, the window investment is attractive.

A.7.4 Internal Rate of Return

One of the problems associated with using the present worth or the annual worth measures of worth is that they depend upon knowing a value for MARR. As mentioned in the introduction to this section, the "proper" value for MARR is a much debated topic and tends to vary from company to company and decision-maker to decision-maker. If the value of MARR changes, the value of PW or AW must be recalculated to determine whether the attractiveness/unattractiveness of an investment has changed.

The internal rate of return (IRR) approach is designed to calculate a rate of return that is "internal" to the project. That is,

if IRR > MARR, the project is attractive;
if IRR < MARR, the project is unattractive; and
if IRR = MARR, the project is indifferent.

Thus, if MARR changes, no new calculations are required. We simply compare the calculated IRR for the project to the new value of MARR, and we have our decision.

The value of IRR is typically determined through a trial and error process. An expression for the present worth of an investment is written without specifying a value for i in the time value of money factors. Then various values of i are substituted until a value is found that sets the present worth (PW) equal to zero. The value of i found in this way is the IRR.

As appealing as the flexibility of this approach is, there are two major drawbacks. First, the iterations required to solve using the trial and error approach to solution can be time consuming. This factor is mitigated by the fact that most spreadsheets and financial calculators are pre-programmed to solve for an IRR value, given a cash flow series. The second (and more serious) drawback to the IRR approach is that some cash flow series have more than one value of IRR—i.e., more than one value of i sets the PW expression to zero. A detailed discussion of this multiple solution issue is beyond the scope of this appendix but can be found in White, et al. [1998], as well as most other economic analysis references. However, it can be shown that, if a cash flow series consists of an initial investment (negative cash flow at $t = 0$) followed by a series of future returns (positive or zero cash flows for all $t > 0$) then a unique IRR exists. If these conditions are not satisfied, a unique IRR is not guaranteed and caution should be exercised in making decisions based on IRR.

The concept of internal rate of return as a measure of investment worth can be generalized as follows:

Measure of Worth: Internal Rate of Return

Description: An interest rate, IRR, is determined which yields a present worth of zero. IRR implicitly assumes the reinvestment of recovered funds at IRR.

Calculation Approach:

$$\text{Find IRR such that PW} = \sum_{t=1}^{n} A_t(P \mid F, IRR, t) = 0.$$

Important Note: Depending upon the cash flow series, multiple IRRs may exist! If the cash flow series consists of an initial investment (net negative cash flow) followed by a series of future returns (net non-negative cash flows), then a unique IRR exists.

Decision Rule: If IRR is unique and IRR ≥MARR, then the investment is attractive.

Example 15
Reconsider the thermal window data of Example 13. If the internal rate of return measure of worth is to be used, is this an attractive investment?

First we note that the cash flow series has a single negative investment, followed by all positive returns; therefore, it has a unique value for IRR. For such a cash flow series it can also be shown that as i increases PW decreases.

From example 11, we know that for i = 15%:
PW =   −10000+2525*(P | A,15%,3)+3840*(P | A,15%,3)*
         (P | F,15%,3)

PW =   −10000+2525*(2.2832)+3840*(2.2832)*
         (0.6575) = $1529.70

Because PW>0, we must increase i to decrease PW toward zero for i = 18%:

PW =   −10000+2525*(P | A,18%,3)+3840*
         (P | A,18%,3)*(P | F,18%,3)

PW =   −10000+2525*(2.1743)+3840*(2.1743)*
         (0.6086) = $571.50

Since PW>0, we must increase i to decrease PW toward zero for i = 20%:

PW =   −10000+2525*(P | A,20%,3)+3840*
         (P | A,20%,3)*(P | F,20%,3)

PW =   −10000+2525*(2.1065)+3840*(2.1065)*
         (0.5787) = −$0.01

Although we could interpolate a value of i for which PW = 0 (rather than −0.01), for practical purposes PW = 0 at i = 20%; therefore, IRR = 20%.

<u>Decision:</u> IRR≥MARR (20%>15%); therefore, the window investment is attractive.

A.7.5 Saving Investment Ratio

Many companies are accustomed to working with benefit cost ratios. An investment measure of worth, which is consistent with the present worth measure and has the form of a benefit cost ratio, is the savings investment ratio (SIR). The SIR decision rule can be derived from the present worth decision rule as follows:

Starting with the PW decision rule

$$PW \geq 0$$

replacing PW with its calculation expression

$$\sum_{t=0}^{n} A_t(P \mid F,i,t) \geq 0$$

which, using the relationship $A_t = R_t - C_t$, can be restated

$$\sum_{t=0}^{n} (R_t - C_t)(P \mid F,i,t) \geq 0$$

which can be algebraically separated into

$$\sum_{t=0}^{n} R_t(P \mid F,i,t) - \sum_{t=0}^{n} C_t(P \mid F,i,t) \geq 0$$

adding the second term to both sides of the inequality

$$\sum_{t=0}^{n} R_t(P \mid F,i,t) \geq \sum_{t=0}^{n} C_t(P \mid F,i,t)$$

dividing both sides of the inequality
by the right side term

$$\frac{\sum\limits_{t=0}^{n} R_t(P\,|\,F,i,t)}{\sum\limits_{t=0}^{n} C_t(P\,|\,F,i,t)} \geq 1$$

which is the decision rule for SIR.

The SIR represents the ratio of the present worth of the revenues to the present worth of the costs. If this ratio exceeds one, the investment is attractive.

The concept of savings investment ratio as a measure of investment worth can be generalized as follows:

Measure of Worth: Savings Investment Ratio

Description: The ratio of the present worth of positive cash flows to the present worth of (the absolute value of) negative cash flows is formed using $i = $ MARR.

$$\text{Calculation Approach: SIR} = \frac{\sum\limits_{t=0}^{n} R_t(P\,|\,F,i,t)}{\sum\limits_{t=0}^{n} C_t(P\,|\,F,i,t)}$$

Decision Rule: If SIR ≥ 1, then the investment is attractive.

Example 16

Reconsider the thermal window data of Example 13. If the savings investment ratio measure of worth is to be used, is this an attractive investment?

From example 13, we know that for $i = 15\%$:

$$\text{SIR} = \frac{\sum\limits_{t=0}^{n} R_t(P\,|\,F,i,t)}{\sum\limits_{t=0}^{n} C_t(P\,|\,F,i,t)}$$

$$SIR = \frac{2525^*(P \mid A, 15\%,3) + 3840^*(P \mid A, 15\%,3)^*(P \mid F, 15\%,3)}{10000}$$

$$SIR = \frac{11529.70}{10000.00} = 1.15297$$

Decision: SIR≥1.0 (1.15297>1.0); therefore, the window investment is attractive.

An important observation regarding the four measures of worth presented to this point (PW, AW, IRR, and SIR) is that they are all consistent and equivalent. In other words, an investment that is attractive under one measure of worth will be attractive under each of the other measures of worth. A review of the decisions determined in Examples 13 through 16 will confirm this observation. Because of their consistency, it is not necessary to calculate more than one measure of investment worth to determine the attractiveness of a project. The rationale for presenting multiple measures which are essentially identical for decision-making is that various individuals and companies may have a preference for one approach over another.

A.7.6 Payback Period

The payback period of an investment is generally taken to mean the number of years required to recover the initial investment through net project returns. The payback period is a popular measure of investment worth and appears in many forms in economic analysis literature and company procedure manuals. Unfortunately, all too frequently, payback period is used inappropriately and leads to decisions which focus exclusively on short term results and ignore time value of money concepts. After presenting a common form of payback period, these shortcomings will be discussed.

Measure of Worth: Payback Period

Description: The number of years required to recover the initial investment by accumulating net project returns is determined.

Calculation Approach:

PBP = the smaller m such that $\sum_{t=1}^{m} A_t \geq C_0$

Decision Rule: If PBP is less than or equal to a predetermined limit (often called a hurdle rate), then the investment is attractive.

Important Note: This form of payback period ignores the time value of money and ignores returns beyond the predetermined limit.

The fact that this approach ignores time value of money concepts is apparent by the fact that no time value of money factors are included in the determination of m. This implicitly assumes that the applicable interest rate to convert future amounts to present amounts is zero. This implies that people are indifferent between $100 today and $100 one year from today, which is an implication that is highly inconsistent with observable behavior.

The short-term focus of the payback period measure of worth can be illustrated using the cash flow diagrams of Figure A-8. Applying the PBP approach above yields a payback period for investment (a) of PBP = 2 (1200>1000 @ t = 2) and a payback period for investment (b) of PBP = 4 (1000300>1000) @ t = 4). If the decision hurdle rate is 3 years (a very common rate), then investment (a) is attractive but investment (b) is not. Hopefully, it is obvious that judging (b) unattractive is not good decision-making, since a $1,000,000 return four years after a $1,000 investment is attractive under almost any value of MARR. In point of fact, the IRR for (b) is 465%, so for any value of MARR less than 465%, investment (b) is attractive.

A.8 ECONOMIC ANALYSIS

A.8.1 Introduction

The general scenario for economic analysis is that a set of investment alternatives are available, and a decision must be made regarding which ones (if any) to accept and which ones (if any) to reject. If the analysis is deterministic, then an assumption is made that cash flow amounts, cash flow timing, and MARR are known with certainty. Frequently, although this assumption does not hold exactly, it is not considered restrictive in terms of potential investment decisions. If, however, the lack of certainty

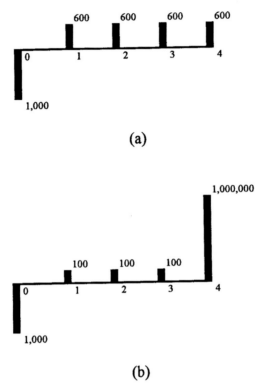

(a)

(b)

Figure A-8. Two investments evaluated using payback period

is a significant issue, then the analysis is stochastic and the assumptions of certainty are relaxed using probability distributions and statistical techniques to conduct the analysis. The remainder of this section deals with deterministic economic analysis, so the assumption of certainty will be assumed to hold. Stochastic techniques are introduced in Section A.9.5.

A.8.2 Deterministic Unconstrained Analysis

Deterministic economic analysis can be further classified into unconstrained deterministic analysis and constrained deterministic analysis. Under unconstrained analysis, all projects within the set available are assumed to be independent. The practical implication of this independence assumption is that an accept/reject decision can be made on each project without regard to the decisions made on other projects. In general this requires that (1) there are sufficient funds available to undertake all proposed projects, (2) there are no mutually exclusive projects, and (3) there are no contingent projects.

A funds restriction creates dependency, since before deciding on a project being evaluated, the evaluator would have to know what decisions had been made on other projects to determine whether sufficient funds were available to undertake the current project. Mutual exclusion creates dependency, since acceptance of one of the mutually exclusive projects precludes acceptance of the others. Contingency creates dependence, since prior to accepting a project, all projects on which it is contingent must be accepted.

If none of the above dependency situations are present and the projects are otherwise independent, then the evaluation of the set of projects is done by evaluating each individual project in turn and accepting the set of projects which were individually judged acceptable. This accept or reject judgment can be made using either the PW, AW, IRR, or SIR measure of worth. The unconstrained decision rules for each or these measures of worth are restated below for convenience:

<u>Unconstrained PW Decision Rule</u>: If PW ≥ 0, then the project is attractive.

<u>Unconstrained AW Decision Rule</u>: If AW ≥ 0, then the project is attractive.

<u>Unconstrained IRR Decision Rule</u>: If IRR is unique and IRR $\geq$ MARR, then the project is attractive.

<u>Unconstrained SIR Decision Rule</u>: If SIR ≥ 1, then the project is attractive.

Example 17

Consider the set of four investment projects whose cash flow diagrams are illustrated in Figure A-9. If MARR is 12%/yr and the analysis is unconstrained, which projects should be accepted?

Using present worth as the measure of worth:

$PW_A = -1000+600*(P \mid A,12\%,4) = -1000+600(3.0373) = \$822.38 \Rightarrow$ Accept A

$PWB = -1300+800*(P \mid A,12\%,4) = -1300+800(3.0373) = \$1129.88 \Rightarrow$ Accept B

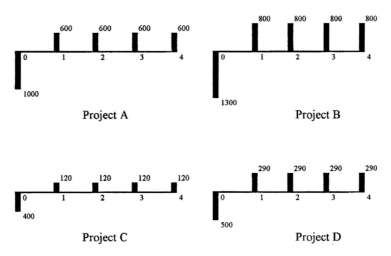

Figure A-9 Four investments projects

$$PWC = -400+120*(P\mid A,12\%,4) = -400+120(3.0373) = -\$35.52 \Rightarrow \text{Reject C}$$

$$PWD = -500+290*(P\mid A,12\%,4) = -500+290(3.0373) = \$380.83 \Rightarrow \text{Accept D}$$

Therefore,
    Accept Projects A, B, and D and reject Project C

A.8.3 Deterministic Constrained Analysis

Constrained analysis is required any time a dependency relationship exists between any of the projects within the set to be analyzed. In general, dependency exists any time (1) there are insufficient funds available to undertake all proposed projects (commonly referred to as capital rationing), (2) there are mutually exclusive projects, or (3) there are contingent projects.

Several approaches have been proposed for selecting the best set of projects from a set of potential projects under constraints. Many of these approaches will select the optimal set of acceptable projects under some conditions (or will select a set that is near optimal). However, only a few approaches are guaranteed to select the optimal set of projects under all conditions. One of these approaches is presented below by way of a continuation of Example 17.

The first steps in the selection process are to specify the cash flow amounts and cash flow timings for each project in the potential project set. Additionally, a value of MARR to be used in the analysis must be specified. These issues have been addressed in previous sections, so further discussion will be omitted here. The next step is to form the set of all possible decision alternatives from the projects. A single decision alternative is a collection of zero, one, or more projects which could be accepted (all others not specified are to be rejected). As an illustration, the possible decision alternatives for the set of projects illustrated in Figure A-9 are listed in Table A-7. As a general rule, there will be 2^n possible decision alternatives generated from a set of n projects. Thus, for the projects of Figure A-9, there are $2^4 = 16$ possible decision alternatives. Since this set represents all possible decisions that could be made,

Table A-7. The decision alternatives from four projects

Accept A only
Accept B only
Accept C only
Accept D only
Accept A and B only
Accept A and C only
Accept A and D only
Accept B and C only
Accept B and D only
Accept C and D only
Accept A, B, and C only
Accept A, B, and D only
Accept A, C, and D only
Accept B, C, and D only
Accept A, B, C, and D (frequently called the do everything alternative)
Accept none (frequently called the do nothing or null alternative)

one, and only one, will be selected as the best (optimal) decision. The set of decision alternatives developed in this way has the properties of being collectively exhaustive (all possible choices are listed) and mutually exclusive (only one will be selected).

The next step in the process is to eliminate decisions from the collectively exhaustive, mutually exclusive set that represent choices which would violate one (or more) of the constraints on the projects. For the projects of Figure A-9, assume the following two constraints exist:

Project B is contingent on Project C, and

A budget limit of $1500 exists on capital expenditures at t = 0.

Based on these constraints the following decision alternatives must be removed from the collectively exhaustive, mutually exclusive set: any combination that includes B but not C (B only, A&B, B&D, A&B&D), any combination not already eliminated whose t = 0 costs exceed $1500 (B&C, A&B&C, A&C&D, B&C&D, A&B&C&D). Thus, from the original set of 16 possible decision alternatives, 9 have been eliminated and need not be evaluated. These results are illustrated in Table A-8. It is frequently the case in practice that a significant percentage of the original collectively exhaustive, mutually exclusive set will be eliminated before measures of worth are calculated.

The next step is to create the cash flow series for the remaining (feasible) decision alternatives. This is a straight forward process and is accomplished by setting a decision alternative's annual cash flow equal to the sum of the annual cash flows (on a year by year basis) of all projects contained in the decision alternative. Table A-9 illustrates the results of this process for the feasible decision alternatives from Table A-8.

The next step is to calculate a measure of worth for each decision alternative. Any of the four consistent measures of worth presented previously (PW, AW, IRR, or SIR but NOT PBP) can be used. The measures are entirely consistent and will lead to the same decision alternative being selected. For illustrative purposes, PW will be calculated for the decision alternatives of Table A-9 assuming MARR = 12%.

Table A-8. The decision alternatives with constraints imposed

Accept A only	OK
Accept B only	infeasible, B contingent on C
Accept C only	OK
Accept D only	OK
Accept A and B only	infeasible, B contingent on C
Accept A and C only	OK
Accept A and D only	OK
Accept B and C only	infeasible, capital rationing
Accept B and D only	infeasible, B contingent on C
Accept C and D only	OK
Accept A, B, and C only	infeasible, capital rationing
Accept A, B, and D only	infeasible, B contingent on C
Accept A, C, and D only	infeasible, capital rationing
Accept B, C, and D only	infeasible, capital rationing
Do Everything	infeasible, capital rationing
null	OK

Table A-9. The decision alternatives cash flows

yr \ Alt	A only	C only	D only	A&C	A&D	C&D	null
0	-1000	-400	-500	-1400	-1500	-900	0
1	600	120	290	720	890	410	0
2	600	120	290	720	890	410	0
3	600	120	290	720	890	410	0
4	600	120	290	720	890	410	0

$PW_A =$ $-1000 + 600^*(P \mid A,12\%,4) = -1000 + 600\ (3.0373)$
              $= \$822.38$

$PW_C =$ $-400 + 120^*(P \mid A,12\%,4) = -400 + 120\ (3.0373)$
              $= -\$35.52$

$PW_D =$ $-500 + 290^*(P \mid A,12\%,4) = -500 + 290\ (3.0373)$
              $= \$380.83$

$PW_{A\&C} =$ $-1400 + 720^*(P \mid A,12\%,4) = -1400 + 720\ (3.0373)$
              $= \$786.86$

$PW_{A\&D} =$ $-1500 + 890^*(P \mid A,12\%,4) = -1500 + 890\ (3.0373)$
              $= \$1203.21$

$PW_{C\&D} =$ $-900 + 410^*(P \mid A,12\%,4) = -900 + 410\ (3.0373)$
              $= \$345.31$

$PW_{null} =$ $-0 + 0^*(P \mid A,12\%,4) = -0 + 0\ (3.0373) = \$0.00$

The decision rules for the various measures of worth under constrained analysis are list below:

Constrained PW Decision Rule: Accept the decision alternative with the highest PW.

Constrained AW Decision Rule: Accept the decision alternative with the highest AW.

Constrained IRR Decision Rule: Accept the decision alternative with the highest IRR.

Constrained SIR Decision Rule: Accept the decision alternative with the highest SIR.

For the example problem, the highest present worth ($1203.21) is associated with accepting projects A and D (rejecting all others). This decision is guaranteed to be optimal (i.e., no feasible combination of projects has a higher PW, AW, IRR, or SIR).

A.8.4 Some Interesting Observations
Regarding Constrained Analysis

Several interesting observations can be made regarding the approach, measures of worth, and decisions associated with constrained analysis. Detailed development of these observations is omitted here but may be found in many engineering economic analysis texts [White, et al., 1998].

- The present worth of a decision alternative is the sum of the present worths of the projects contained within the alternative. (From above, $PW_{A\&D} = PW_A + PW_D$).

- The annual worth of a decision alternative is the sum of the annual worths of the projects contained within the alternative.

- The internal rate of return of a decision alternative is NOT the sum of internal rates of returns of the projects contained within the alternative. The IRR for the decision alternative must be calculated by the trial and error process of finding the value of i that sets the PW of the decision alternative to zero.

- The savings investment ratio of a decision alternative is NOT the sum of the savings investment ratios of the projects contained within the alternative. The SIR for the decision alternative must be calculated from the cash flows of the decision alternative.

- A common, but flawed, procedure for selecting the projects to accept from the set of potential projects involves ranking the projects (not decision alternatives) in preferred order based on a measure of worth calculated for the project (e.g., decreasing project PW) and then accepting projects as far down the list as funds allow. While this procedure will select the optimal set under some conditions (e.g., it works well if the initial investments of all projects are small relative to the capital budget limit), it is not guaranteed to select the optimal set under all conditions. The procedure outlined above will select the optimal set under all conditions.

- Table A-10 illustrates that the number of decision alternatives in the collectively exhaustive, mutually exclusive set can grow prohibitively large as the number of potential projects increases. The mitigating factor in this combinatorial growth problem is that in most practical situations a high percentage of the possible decision alternatives are infeasible and do not require evaluation.

**Table A-10. The number of decision alternatives
as a function of the number of projects**

Number of Projects	Number of Decision Alternatives
1	2
2	4
3	8
4	16
5	32
6	64
7	128
8	256
0	512
10	1,024
15	32,768
20	1,048,576
25	33,554,432

A.8.5 The Planning Horizon Issue

When comparing projects, it is important to compare the costs and benefits over a common period of time. The intuitive sense of fairness here is based upon the recognition that most consumers expect an investment that generates savings over a longer period of time to cost more than an investment that generates savings over a shorter period of time. To facilitate a fair, comparable evaluation, a common period of time over which to conduct the evaluation is required. This period of time is referred to as the planning horizon. The planning horizon issue arises when at least one project has cash flows defined over a life which is greater than or less than the life of at least one other project. This situation did not occur in Example 17 of the previous section, since all projects had 4 year lives.

There are four common approaches to establishing a planning horizon for evaluating decision alternatives. These are (1) shortest life, (2) longest life, (3) least common multiple of lives, and (4) standard. The shortest life planning horizon is established by selecting the project with the shortest life and setting this life as the planning horizon. A significant issue in this approach is how to value the remaining cash flows for projects whose lives are truncated. The typical approach to this valuation is to estimate the value of the remaining cash flows as the salvage value (market value) of the investment at that point in its life.

Example 18

Determine the shortest life planning horizon for projects A, B, C with lives 3, 5, and 6 years, respectively.

The shortest life planning horizon is 3 years based on Project A. A salvage value must be established at $t = 3$ for B's cash flows in years 4 and 5. A salvage value must be established at $t = 3$ for C's cash flows in years 4, 5, and 6.

The longest life planning horizon is established by selecting the project with the longest life and setting this life as the planning horizon. The significant issue in this approach is how to handle projects whose cash flows don't extend this long. The typical resolution for this problem is to assume that shorter projects are repeated consecutively (end-to-end) until one of the repetitions extends at least as far as the planning horizon. The assumption of project repeatability deserves careful consideration, since in some cases it is reasonable and in others it may be quite unreasonable. The reasonableness of the assumption is largely a function of the type of investment and the rate of innovation occurring within the investment's field. (For example, assuming repeatability of investments in high technology equipment is frequently ill advised, since the field is advancing rapidly.) If in repeating a project's cash flows, the last repetition's cash flows extend beyond the planning horizon, then the truncated cash flows (those that extend beyond the planning horizon) must be assigned a salvage value as above.

Example 19

Determine the longest life planning horizon for projects A, B, C with lives 3, 5, and 6 years, respectively.

The longest life planning horizon is 6 years based on Project C. Project A must be repeated twice, with the second repetition ending at year 6, so no termination of cash flows is required. Project B's second repetition extends to year 10; therefore, a salvage value at t = 6 must be established for B's repeated cash flows in years 7, 8, 9, and 10.

An approach that eliminates the truncation salvage value issue from the planning horizon question is the least common multiple approach. The least common multiple planning horizon is set by determining the smallest number of years at which repetitions of all projects would terminate simultaneously. The least common multiple for a set of numbers (lives) can be determined mathematically using algebra. Discussion of this approach is beyond the scope of this appendix. For a small number of projects, the value can be determined with trial and error by examining multiples of the longest life project.

Example 20

Determine the least common multiple planning horizon for projects A, B, C with lives 3, 5, and 6 years, respectively.

The least common multiple of 3, 5, and 6 is 30. This can be obtained by trial and error, starting with the longest project life (6) as follows:

1st trial: 6*1 = 6; 6 is a multiple of 3 but not 5; reject 6 and proceed

2nd trial: 6*2 = 12; 12 is a multiple of 3 but not 5; reject 12 and proceed

3rd trial: 6*3 = 18; 18 is a multiple of 3 but not 5; reject 18 and proceed

4th trial: 6*4 = 24; 24 is a multiple of 3 but not 5; reject 24 and proceed

5th trial: 6*5 = 30; 30 is a multiple of 3 and 5; accept 30 and stop

Under a 30-year planning horizon, A's cash flows are repeated 10 times, B's 6 times, and C's 5 times. No truncation is required.

The standard planning horizon approach uses a planning horizon that is independent of the projects being evaluated. Typically, this type of planning horizon is based on company policies or practices. The standard horizon may require repetition and/or truncation, depending upon the set of projects being evaluated.

Example 21
Determine the impact of a 5 year standard planning horizon on projects A, B, C with lives 3, 5, and 6 years, respectively.

With a 5-year planning horizon:

Project A must be repeated one time, with the second repetition truncated by one year.

Project B is a 5 year project and does not require repetition or truncation.

Project C must be truncated by one year.

There is no single accepted approach to resolving the planning horizon issue. Companies and individuals generally use one of the approaches outlined above. The decision of which to use in a particular analysis is generally a function of company practice, consideration of the reasonableness of the project repeatability assumption, and the availability of salvage value estimates at truncation points.

A.9 SPECIAL PROBLEMS

A.9.1 Introduction
The preceding sections of this appendix outline an approach for conducting deterministic economic analysis of investment opportunities. Adherence to the concepts and methods presented will lead to sound investment decisions with respect to time value of money principles. This section addresses several topics that are of special interest in some analysis situations.

A.9.2 Interpolating Interest Tables
All of the examples previously presented in this appendix conveniently used interest rates whose time value of money factors were tabulated in Appendix 4A. How does one proceed if non-tabulated time value of money factors are needed? There are two viable approaches, calculation

of the exact values and interpolation. The best and theoretically correct approach is to calculate the exact values of needed factors, based on the formulas in Table A-6.

Example 22
    Determine the exact value for (F I P,13%,7).

From Table A-6,

$$(F \mid P,i,n) = (1+i)^n = (1+.13)^7 = 2.3526$$

Interpolation is often used instead of calculation of exact values because, with practice, interpolated values can be calculated quickly. Interpolated values are not "exact," but for most practical problems they are "close enough," particularly if the range of interpolation is kept as narrow as possible. Interpolation of some factors, for instance (P I A,i,n), also tends to be less error prone than the exact calculation due of simpler mathematical operations.

Interpolation involves determining an unknown time value of money factor using two known values that bracket the value of interest. An assumption is made that the values of the time value of money factor vary linearly between the known values. Ratios are then used to estimate the unknown value. The example below illustrates the process.

Example 23
    Determine an interpolated value for (F I P,13%,7).

The narrowest range of interest rates that bracket 13%, and for which time value of money factor tables are provided in Appendix 4A, is 12% to 15%.

    The values necessary for this interpolation are

i values	(F«F,i%,7)
12%	2.2107
13%	(F I P,13%,7)
15%	2.6600

The interpolation proceeds by setting up ratios and solving for the unknown value, (F | P,13%,7), as follows:

$$\frac{\text{change between rows 2 \& 1 of left column}}{\text{change between rows 3 \& 1 of left column}} =$$

$$\frac{\text{change between rows 2 \& 1 of right column}}{\text{change between rows 3 \& 1 of right column}}$$

$$\frac{0.13 - 0.12}{0.15 - 0.12} = \frac{(F \mid P,13\%,7) - 2.2107}{2.6600 - 2,2107}$$

$$\frac{0.01}{0.03} = \frac{(F \mid P,13\%,7) - 2.2107}{0.4493}$$

$$0.1498 = (F \mid P,13\%,7) - 2.2107$$

$$(F \mid P,13\%,7) = 2.3605$$

The interpolated value for (F | P,13%,7), 2.3605, differs from the exact value, 2.3526, by 0.0079. This would imply a $7.90 difference in present worth for every thousand dollars of return at t = 7. The relative importance of this interpolation error can be judged only in the context of a specific problem.

A.9.3 Non-Annual Interest Compounding

Many practical economic analysis problems involve interest that is not compounded annually. It is common practice to express a non-annually compounded interest rate as follows:

12% per year compounded monthly or 12%/yr/mo.

When expressed in this form, 12%/yr/mo is known as the nominal annual interest rate. The techniques covered in this appendix up to this point can not be used directly to solve an economic analysis problem of this type because the interest period (per year) and compounding period (monthly) are not the same. Two approaches can be used to solve problems of this type. One approach involves determining a period interest rate; the other involves determining an effective interest rate.

To solve this type of problem using a period interest rate approach, we must define the period interest rate:

$$\text{Period Interest Rate} = \frac{\text{Nominal Annual Interest Rate}}{\text{Number of Interest Periods per Year}}$$

In our example,

$$\text{Period Interest Rate} = \frac{12\%/\text{yr}/\text{mo}}{12 \text{ mo}/\text{yr}} = 1\%/\text{mo}/\text{mo}$$

Because the interest period and the compounding period are now the same, the time value of money factors in Appendix 4A can be applied directly. Note, however, that the number of interest periods (n) must be adjusted to match the new frequency.

Example 24

$2,000 is invested in an account which pays 12% per year compounded monthly. What is the balance in the account after 3 years?

Nominal Annual Interest Rate = 12%/yr/mo

$$\text{Period Interest Rate} = \frac{12\%/\text{yr}/\text{mo}}{12 \text{ mo}/\text{yr}} = 1\%/\text{mo}/\text{mo}$$

Number of Interest Periods = 3 years × 12 mo/yr = 36 interest periods (months)

$$F = P\,(F\,|\,P,i,n) = \$2,000\,(F\,|\,P,1,36) = \$2,000\,(1.4308) = \$2,861.60$$

Example 25

What are the monthly payments on a 5-year car loan of $12,500 at 6% per year compounded monthly?

Nominal Annual Interest Rate = 6%/yr/mo

$$\text{Period Interest Rate} = \frac{6\%/\text{yr}/\text{mo}}{12 \text{ mo}/\text{yr}} = 0.5\%/\text{mo}/\text{mo}$$

Number of Interest Periods = 5 years × 12 mo/yr = 60 interest periods

$$A = P \, (A \mid P,i,n) = \$12,500 \, (A \mid P,0.5,60) = \$12,500 \, (0.0193) = \$241.25$$

To solve this type of problem using an effective interest rate approach, we must define the effective interest rate. The effective annual interest rate is the annualized interest rate that would yield results equivalent to the period interest rate as previously calculated. However, the effective annual interest rate approach should not be used if the cash flows are more frequent than annual (e.g., monthly). In general, the interest rate for time value of money factors should match the frequency of the cash flows. (For example, if the cash flows are monthly, use the period interest rate approach with monthly periods.)

As an example of the calculation of an effective interest rate, assume that the nominal interest rate is 12%/yr/qtr; therefore, the period interest rate is 3%/qtr/qtr. One dollar invested for 1 year at 3%/qtr/qtr would have a future worth as calculated:

$$F = P \, (F \mid P,i,n) = \$1 \, (F \mid P,3,4) = \$1 \, (1.03)^4$$
$$= \$1 \, (1.1255) = \$1.1255$$

To get this same value in 1 year with an annual rate, the annual rate would have to be of 12.55%/yr/yr. This value is called the effective annual interest rate. The effective annual interest rate is given by $(1.03)^4 - 1 = 0.1255$ or 12.55%.

The general equation for the Effective Annual Interest Rate is:

Effective Annual Interest Rate = $(1 + (r/m))^m - 1$
where: r = nominal annual interest rate
m = number of interest periods per year

Example 26

What is the effective annual interest rate if the nominal rate is 12%/yr compounded monthly?

nominal annual interest rate = 12%/yr/mo

period interest rate = 1%/mo/mo

effective annual interest rate = $(1+0.12/12)^{12} - 1 = 0.1268$ or 12.68%

A.9.4 Economic Analysis Under Inflation

Inflation is characterized by a decrease in the purchasing power of money that is caused by an increase in general price levels of goods and services without an accompanying increase in value. Inflationary pressure is created when more dollars are put into an economy with no accompanying increase in goods and services. In other words, printing more money without an increase in economic output generates inflation. A complete treatment of inflation is beyond the scope of this appendix. A good summary can be found in Sullivan and Bontadelli [1980].

When consideration of inflation is introduced into economic analysis, future cash flows can be stated in terms of either constant-worth dollars or then-current dollars. Then-current cash flows are expressed in terms of the face amount of dollars (actual number of dollars) that will change hands when the cash flow occurs. Alternatively, constant-worth cash flows are expressed in terms of the purchasing power of dollars relative to a fixed point in time known as the base period.

Example 27

For the next 4 years, a family anticipates buying $1000 worth of groceries each year. If inflation is expected to be 3%/yr, what are the then-current cash flows required to purchase the groceries?

To buy the groceries, the family will need to take the following face amount of dollars to the store. We will somewhat artificially assume that the family only shops once per year, buys the same set of items each year, and that the first trip to the store will be one year from today.

Year 1: dollars required $1000.00*(1.03) = $1030.00

Year 2: dollars required $1030.00*(1.03) = $1060.90

Year 3: dollars required $1060.90*(1.03) = $1092.73

Year 4: dollars required $1092.73*(1.03) = $1125.51

What are the constant-worth cash flows, if today's dollars are used as the base year?

The constant worth dollars are inflation free dollars; therefore, the $1000 of groceries costs $1000 each year.

Year 1: $1000.00

Year 2: $1000.00

Year 3: $1000.00

Year 4: $1000.00

The key to proper economic analysis under inflation is to base the value of MARR on the types of cash flows. If the cash flows contain inflation, then the value of MARR should also be adjusted for inflation. Alternatively, if the cash flows do not contain inflation, then the value of MARR should be inflation-free. When MARR does not contain an adjustment for inflation, it is referred to as a real value for MARR. If it contains an inflation adjustment, it is referred to as a combined value for MARR. The relationship between inflation rate, the real value of MARR, and the combined value of MARR is given by:

$$1 + MARR_{COMBINED}$$
$$= (1 + \text{inflation rate}) * (1 + MARR_{REAL})$$

Example 28

If the inflation rate is 3%/yr and the real value of MARR is 15%/yr, what is the combined value of MARR?

$$1 + MARR_{COMBINED}$$
$$= (1 + \text{inflation rate}) * (1 + MARR_{REAL})$$
$$1 + MARR_{COMBINED} = (1 + 0.03) * (1 + 0.15)$$
$$1 + MARR_{COMBINED} = (1.03) * (1.15)$$
$$1 + MARR_{COMBINED} = 1.1845$$
$$MARR_{COMBINED} = 1.1845 - 1 = 0.1845 = 18.45\%$$

If the cash flows of a project are stated in terms of then-current dollars, the appropriate value of MARR is the combined value of MARR. Analysis done in this way is referred to as then current analysis. If the cash flows of a project are stated in terms of constant-worth dollars, the appropriate value of MARR is the real value of MARR. Analysis done in this way is referred to as then constant worth analysis.

Example 29

Using the cash flows of Examples 27 and interest rates of Example 28, determine the present worth of the grocery purchases using a constant worth analysis.

Constant worth analysis requires constant worth cash flows and the real value of MARR.

$$PW = 1000 * (P \mid A, 15\%, 4)$$
$$= 1000 * (2.8550) = \$2855.00$$

Example 30

Using the cash flows of Examples 27 and interest rates of Example 28, determine the present worth of the grocery purchases using a then current analysis.

Then current analysis requires then current cash flows and the combined value of MARR.

$$PW = 1030.00 * (P \mid F, 18.45\%, 1) + 1060.90 * (P \mid F, 18.45\%, 2) + 1092.73 * (P \mid F, 18.45\%, 3) + 1125.51 * (P \mid F, 18.45\%, 4)$$

$$PW = 1030.00 * (0.8442) + 1060.90 * (0.7127) + 1092.73 * (0.6017) + 1125.51 * (0.5080)$$

$$PW = 869.53 + 756.10 + 657.50 + 571.76 = 2854.89$$

The notable result of Examples 29 and 30 is that the present worths determined by the constant-worth approach ($2855.00) and the then-current approach ($2854.89) are equal. (The $0.11 difference is due to rounding.) This result is often unexpected but is mathematically sound. The important conclusion is that if care is taken to appropriately match the cash flows and value of MARR, the level of general price inflation is not a determining factor in the acceptability of projects. To make this important result hold, inflation must either (1) be included in both the cash flows, and MARR (the then-current approach) or (2) be included in neither the cash flows nor MARR (the constant-worth approach).

A.9.5 Sensitivity Analysis and Risk Analysis

Often times the certainty assumptions associated with deterministic analysis are questionable. These certainty assumptions include certain knowledge regarding amounts and timing of cash flows, as well as certain knowledge of MARR. Relaxing these assumptions requires the use of sensitivity analysis and risk analysis techniques.

Initial sensitivity analyses are usually conducted on the optimal decision alternative (or top two or three) on a single factor basis. Single factor sensitivity analysis involves holding all cost factors except one constant while varying the remaining cost factor through a range of percentage changes. The effect of cost factor changes on the measure of worth is observed, to determine whether the alternative remains attractive under the evaluated changes and to determine which cost factor effects the measure of worth the most.

Example 31

Conduct a sensitivity analysis of the optimal decision resulting from the constrained analysis of the data in Example 17. The sensitivity analysis should explore the sensitivity of present worth to changes in annual revenue over the range -10% to $+10\%$.

The PW of the optimal decision (Accept A & D only) was determined in Section A.8.3 to be:

$$PW_{A\&D} = -1500 + 890*(P\,|\,A,12\%,4) = -1500 + 890\,(3.0373) = \$1203.21$$

If annual revenue decreases 10%, it becomes $890 - 0.10*890 = 801$ and PW becomes

$$PW_{A\&D} = -1500 + 801*(P\,|\,A,12\%,4) = -1500 + 801\,(3.0373) = \$932.88$$

If annual revenue increases 10%, it becomes $890 + 0.10*890 = 979$ and PW becomes

$$PW_{A\&D} = -1500 + 979*(P\,|\,A,12\%,4) = -1500 + 979\,(3.0373) = \$1473.52$$

The sensitivity of PW to changes in annual revenue over the range -10% to $+10\%$ is $+\$540.64$ (from $\$932.88$ to $\$1473.52$).

Example 32

Repeat Example 31, exploring the sensitivity of present worth to changes in initial cost over the range -10% to $+10\%$.

The PW of the optimal decision (Accept A & D only) was determined in Section A.8.3 to be:

$PW_{A\&D}$ = $-$ 1500 + 890*(P I A,12%,4) = $-$ 1500 + 890 (3.0373) = $1203.21

If initial cost decreases 10% it becomes 1500 – 0.10*1500 = 1350 and PW becomes

$PW_{A\&D}$ = $-$ 1350 + 890*(P I A,12%,4) = $-$ 1350 + 890 (3.0373) = $1353.20

If initial cost increases 10% it becomes 1500 + 0.10*1500 = 1650 and PW becomes

$PW_{A\&D}$ = $-$ 1650 + 890*(P I A,12%,4) = $-$ 1500 + 890 (3.0373) = $1053.20

The sensitivity of PW to changes in initial cost over the range – 10% to +10% is – $300.00 (from $1353.20 to $1053.20).

Example 33

Repeat Example 31 exploring the sensitivity of the present worth to changes in MARR over the range – 10% to +10%.

The PW of the optimal decision (Accept A & D only) was determined in Section A.8.3 to be:

$PW_{A\&D}$ = $-$ 1500 + 890*(P I A,12%,4) = $-$ 1500 + 890 (3.0373) = $1203.21

If MARR decreases 10% it becomes 12% – 0.10*12% = 10.8% and PW becomes

$PW_{A\&D}$ = $-$ 1500 + 890*(P I A,10.8%,4) = $-$ 1500 + 890 (3.1157) = $1272.97

If MARR increases 10% it becomes 12% + 0.10*12% = 13.2% and PW becomes

$PW_{A\&D}$ = $-$ 1500 + 890*(P I A,13.2%,4) = $-$ 1500 + 890 (2.9622) = $1136.36

The sensitivity of PW to changes in MARR over the range – 10% to +10% is – $136.61 (from $1272.97 to $1136.36).

The sensitivity data from Examples 31, 32, and 33 are summarized in Table A-11. A review of the table reveals that the decision alternative A&D remains attractive (PW ≥0) within the range of 10% changes in annual revenues, initial cost, and MARR. An appealing way to summarize single factor sensitivity data is to use a "spider" graph. A spider graph plots the PW values determined in the examples and connects them with lines, one line for each factor evaluated. Figure A-10 illustrates the spider graph for the data of Table A-11. On this graph, lines with large positive or negative slopes (angle relative to horizontal regardless of whether it is increasing or decreasing) indicate factors to which the present value measure of worth is sensitive. Figure A-10 shows that PW is least sensitive to changes in MARR (the MARR line is the most nearly horizontal) and most sensitive to changes in annual revenue. (The annual revenue line has the steepest slope.) Additional sensitivity could be explored in a similar manner.

When single factor sensitivity analysis is inadequate to assess the questions that surround the certainty assumptions of a deterministic analysis, risk analysis techniques can be employed. One approach to risk

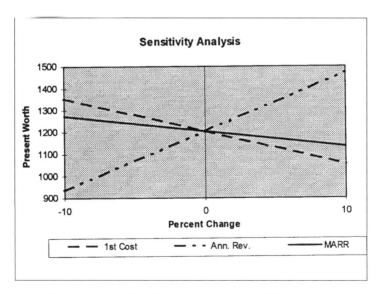

Figure A-10. Sensitivity analysis "spider" graph

Table A-11 Sensitivity analysis data table

Factor/ Percent Change	- 10%	Base	+ 10%
1st Cost	1353.20	1203.21	1053.20
Annual Revenue	932.88	1203.21	1473.52
MARR	1272.97	1203.21	1136.36

analysis is the application of probabilistic and statistical concepts to economic analysis. These techniques require information regarding the possible values that uncertain quantities may take on, as well as estimates of the probability that the various values will occur. A detailed treatment of this topic is beyond the scope of this appendix. A good discussion of this subject can be found in Park and Sharp-Bette [1990].

A second approach to risk analysis in economic analysis is through the use of simulation techniques and simulation software. Simulation involves using a computer simulation program to sample possible values for the uncertain quantities in an economic analysis and then calculating the measure of worth. This process is repeated many times using different samples each time. After many samples have been taken, probability statements regarding the measure of worth may be made. A good discussion of this subject can be found in Park and Sharp-Bette [1990].

A.10 SUMMARY AND ADDITIONAL
EXAMPLE APPLICATIONS

In this appendix a coherent, consistent approach to economic analysis of capital investments (energy related or other) has been presented. To conclude the appendix, this section provides several additional examples to illustrate the use of time value of money concepts for energy related problems. Additional example applications, as well as a more in depth presentation of conceptual details, can be found in the references listed at the end of the appendix. These references are by no means exclusive; many other excellent presentations of the subject matter are also available. Adherence to the concepts and methods presented here and in the references will lead to sound investment decisions with respect to time value of money principles.

Example 34

In Section A.3.3 an example involving the evaluation of a baseboard heating and window air conditioner versus a heat pump was introduced to illustrate cash flow diagramming (Figure A-2). A summary of the differential costs is repeated here for convenience:

- The heat pump costs $1500 more than the baseboard system.

- The heat pump saves $380 annually in electricity costs.

- The heat pump has a $50 higher annual maintenance costs.

- The heat pump has a $150 higher salvage value at the end of 15 years.

- The heat pump requires $200 more in replacement maintenance at the end of year 8.

If MARR is 18%, is the additional investment in the heat pump attractive?

Using present worth as the measure of worth:

$$PW = -1500 + 380*(P \mid A,18\%,15) - 50*(P \mid A,18\%,15) + 150*(P \mid F,18\%,15) - 200*(P \mid F,18\%,8)$$

$$PW = -1500 + 380*(5.0916) - 50*(5.0916) + 150*(0.0835) - 200*(0.2660)$$

$$PW = -1500.00 + 1934.81 - 254.58 + 12.53 - 53.20 = \$139.56$$

Decision: PW≥0 ($139.56>0.0); therefore, the additional investment for the heat pump is attractive.

Example 35

A homeowner needs to decide whether to install R-11 or R-19 insulation in the attic of her home. The R-19 insulation costs $150 more to install and will save approximately 400 kWh per year. If the planning horizon is 20 years and electricity costs $0.08/kWh, is the additional investment attractive at MARR of 10%?

At $0.08/kWh, the annual savings are: 400 kWh * $0.08/kWh = $32.00

Using present worth as the measure of worth:

PW = – 150 + 32*(P I A,10%,20)

PW = – 150 + 32*(8.5136) = – 150 + 272.44 = $122.44

Decision: PW≥0 ($122.44>0.0); therefore, the R-19 insulation is attractive.

Example 36

The homeowner from Example 35 can install R-30 insulation in the attic of her home for $200 more than the R-19 insulation. The R-30 will save approximately 250 kWh per year over the R-19 insulation. Is the additional investment attractive?

Assuming the same MARR, electricity cost, and planning horizon, the additional annual savings are: 250 kWh * $0.08/kWh = $20.00

Using present worth as the measure of worth:

PW = – 200 + 20*(P I A,10%,20)

PW = – 200 + 20*(8.5136) = – 200 + 170.27 = – $29.73

Decision: PW<0 (-$29.73<0.0); therefore, the R-30 insulation is not attractive.

Example 37

An economizer costs $20,000 and will last 10 years. It will generate savings of $3,500 per year with maintenance costs of $500 per year. If M1ARR is 10%, is the economizer an attractive investment.

Using present worth as the measure of worth:

PW = − 20000 + 3500*(P | A,10%,10) − 500*(P | A,10%,10)

PW = − 20000 + 3500*(6.1446) − 500*(6.1446)

PW = − 20000.00 + 21506.10 − 3072.30 = − $1566.20

Decision: PW<0 (-$1566.20<0.0); therefore, the economizer is not attractive.

Example 38

If the economizer from Example 37 has a salvage value of $5000 at the end of 10 years, is the investment attractive?

Using present worth as the measure of worth:

PW = − 20000 + 3500*(P | A,10%,10)
    − 500*(P | A,10%,10) + 5000*(P | F,10%,10)

PW = − 20000 + 3500*(6.1446) − 500*(6.1446) + 5000*(0.3855)

PW = − 20000.00 + 21506.10 − 3072.30 + 1927.50 = $361.30

Decision: PW≥0 ($361.30≥0.0); therefore, the economizer is now attractive.

A.11 References

Brown, R.J. and R.R. Yanuck, 1980, *Life Cycle Costing: A Practical Guide for Energy Managers*, The Fairmont Press, Inc., Atlanta, GA.

Fuller, S.K. and S.R. Petersen, 1994, NISTIR 5165: Life-Cycle Costing Workshop for Energy Conservation in Buildings: Student Manual, U.S. Department of Commerce, Office of Applied Economics, Gaithersburg, MD.

Fuller, S.K. and S.R. Petersen, 1995, NIST Handbook 135: Life-Cycle Costing Manual for the Federal Energy Management Program, National Technical Information Service, Springfield, VA.

Park, C.S. and G.P. Sharp-Bette, 1990, *Advanced Engineering Economics*, John Wiley & Sons, New York, NY.

Sullivan, W.G. and J.A. Bontadelli, 1980, "The Industrial Engineer and Inflation," *Industrial Engineering*, Vol. 12, No. 3, 24-33.

Thuesen, G.J. and W.J. Fabrycky, 1993, *Engineering Economy, 8th Edition*, Prentice Hall, Englewood Cliffs, NJ.

White, J.A., K.E. Case, D.B. Pratt, and M.H. Agee, 1998, *Principles of Engineering Economic Analysis, 3rd Edition*, John Wiley & Sons, New York, NY.

Time Value of Money Factors—Discrete Compounding
i = 1%

n	Single Sums		Uniform Series				Gradient Series	
	To Find F Given P (F|P,i%,n)	To Find P Given F (P|F,i%,n)	To Find F Given A (F|A,i%,n)	To Find A Given F (A|F,i%,n)	To Find P Given A (P|A,i%,n)	To Find A Given P (A|P,i%,n)	To Find P Given G (P|G,i%,n)	To Find A Given G (A|G,i%,n)
1	1.0100	0.9901	1.0000	1.0000	0.9901	1.0100	0.0000	0.0000
2	1.0201	0.9803	2.0100	0.4975	1.9704	0.5075	0.9803	0.4975
3	1.0303	0.9706	3.0301	0.3300	2.9410	0.3400	2.9215	0.9934
4	1.0406	0.9610	4.0604	0.2463	3.9020	0.2563	5.8044	1.4876
5	1.0510	0.9515	5.1010	0.1960	4.8534	0.2060	9.6103	1.9801
6	1.0615	0.9420	6.1520	0.1625	5.7955	0.1725	14.3205	2.4710
7	1.0721	0.9327	7.2135	0.1386	6.7282	0.1486	19.9168	2.9602
8	1.0829	0.9235	8.2857	0.1207	7.6517	0.1307	26.3812	3.4478
9	1.0937	0.9143	9.3685	0.1067	8.5660	0.1167	33.6959	3.9337
10	1.1046	0.9053	10.4622	0.0956	9.4713	0.1056	41.8435	4.4179
11	1.1157	0.8963	11.5668	0.0865	10.3676	0.0965	50.8067	4.9005
12	1.1268	0.8874	12.6825	0.0788	11.2551	0.0888	60.5687	5.3815
13	1.1381	0.8787	13.8093	0.0724	12.1337	0.0824	71.1126	5.8607
14	1.1495	0.8700	14.9474	0.0669	13.0037	0.0769	82.4221	6.3384
15	1.1610	0.8613	16.0969	0.0621	13.8651	0.0721	94.4810	6.8143
16	1.1726	0.8528	17.2579	0.0579	14.7179	0.0679	107.2734	7.2886
17	1.1843	0.8444	18.4304	0.0543	15.5623	0.0643	120.7834	7.7613
18	1.1961	0.8360	19.6147	0.0510	16.3983	0.0610	134.9957	8.2323
19	1.2081	0.8277	20.8109	0.0481	17.2260	0.0581	149.8950	8.7017
20	1.2202	0.8195	22.0190	0.0454	18.0456	0.0554	165.4664	9.1694
21	1.2324	0.8114	23.2392	0.0430	18.8570	0.0530	181.6950	9.6354
22	1.2447	0.8034	24.4716	0.0409	19.6604	0.0509	198.5663	10.0998
23	1.2572	0.7954	25.7163	0.0389	20.4558	0.0489	216.0660	10.5626
24	1.2697	0.7876	26.9735	0.0371	21.2434	0.0471	234.1800	11.0237
25	1.2824	0.7798	28.2432	0.0354	22.0232	0.0454	252.8945	11.4831
26	1.2953	0.7720	29.5256	0.0339	22.7952	0.0439	272.1957	11.9409
27	1.3082	0.7644	30.8209	0.0324	23.5596	0.0424	292.0702	12.3971
28	1.3213	0.7568	32.1291	0.0311	24.3164	0.0411	312.5047	12.8516
29	1.3345	0.7493	33.4504	0.0299	25.0658	0.0399	333.4863	13.3044
30	1.3478	0.7419	34.7849	0.0287	25.8077	0.0387	355.0021	13.7557
36	1.4308	0.6989	43.0769	0.0232	30.1075	0.0332	494.6207	16.4285
42	1.5188	0.6584	51.8790	0.0193	34.1581	0.0293	650.4514	19.0424
48	1.6122	0.6203	61.2226	0.0163	37.9740	0.0263	820.1460	21.5976
54	1.7114	0.5843	71.1410	0.0141	41.5687	0.0241	1.002E+03	24.0945
60	1.8167	0.5504	81.6697	0.0122	44.9550	0.0222	1.193E+03	26.5333
66	1.9285	0.5185	92.8460	0.0108	48.1452	0.0208	1.392E+03	28.9146
72	2.0471	0.4885	104.7099	9.550E-03	51.1504	0.0196	1.598E+03	31.2386
120	3.3004	0.3030	230.0387	4.347E-03	69.7005	0.0143	3.334E+03	47.8349
180	5.9958	0.1668	499.5802	2.002E-03	83.3217	0.0120	5.330E+03	63.9697
360	35.9496	0.0278	3.495E+03	2.861E-04	97.2183	0.0103	8.720E+03	89.6995

Time Value of Money Factors—Discrete Compounding
i = 2%

	Single Sums		Uniform Series				Gradient Series	
	To Find F Given P	To Find P Given F	To Find F Given A	To Find A Given F	To Find P Given A	To Find A Given P	To Find P Given G	To Find A Given G
n	(F|P,i%,n)	(P|F,i%,n)	(F|A,i%,n)	(A|F,i%,n)	(P|A,i%,n)	(A|P,i%,n)	(P|G,i%,n)	(A|G,i%,n)
1	1.0200	0.9804	1.0000	1.0000	0.9804	1.0200	0.0000	0.0000
2	1.0404	0.9612	2.0200	0.4950	1.9416	0.5150	0.9612	0.4950
3	1.0612	0.9423	3.0604	0.3268	2.8839	0.3468	2.8458	0.9868
4	1.0824	0.9238	4.1216	0.2426	3.8077	0.2626	5.6173	1.4752
5	1.1041	0.9057	5.2040	0.1922	4.7135	0.2122	9.2403	1.9604
6	1.1262	0.8880	6.3081	0.1585	5.6014	0.1785	13.6801	2.4423
7	1.1487	0.8706	7.4343	0.1345	6.4720	0.1545	18.9035	2.9208
8	1.1717	0.8535	8.5830	0.1165	7.3255	0.1365	24.8779	3.3961
9	1.1951	0.8368	9.7546	0.1025	8.1622	0.1225	31.5720	3.8681
10	1.2190	0.8203	10.9497	0.0913	8.9826	0.1113	38.9551	4.3367
11	1.2434	0.8043	12.1687	0.0822	9.7868	0.1022	46.9977	4.8021
12	1.2682	0.7885	13.4121	0.0746	10.5753	0.0946	55.6712	5.2642
13	1.2936	0.7730	14.6803	0.0681	11.3484	0.0881	64.9475	5.7231
14	1.3195	0.7579	15.9739	0.0626	12.1062	0.0826	74.7999	6.1786
15	1.3459	0.7430	17.2934	0.0578	12.8493	0.0778	85.2021	6.6309
16	1.3728	0.7284	18.6393	0.0537	13.5777	0.0737	96.1288	7.0799
17	1.4002	0.7142	20.0121	0.0500	14.2919	0.0700	107.5554	7.5256
18	1.4282	0.7002	21.4123	0.0467	14.9920	0.0667	119.4581	7.9681
19	1.4568	0.6864	22.8406	0.0438	15.6785	0.0638	131.8139	8.4073
20	1.4859	0.6730	24.2974	0.0412	16.3514	0.0612	144.6003	8.8433
21	1.5157	0.6598	25.7833	0.0388	17.0112	0.0588	157.7959	9.2760
22	1.5460	0.6468	27.2990	0.0366	17.6580	0.0566	171.3795	9.7055
23	1.5769	0.6342	28.8450	0.0347	18.2922	0.0547	185.3309	10.1317
24	1.6084	0.6217	30.4219	0.0329	18.9139	0.0529	199.6305	10.5547
25	1.6406	0.6095	32.0303	0.0312	19.5235	0.0512	214.2592	10.9745
26	1.6734	0.5976	33.6709	0.0297	20.1210	0.0497	229.1987	11.3910
27	1.7069	0.5859	35.3443	0.0283	20.7069	0.0483	244.4311	11.8043
28	1.7410	0.5744	37.0512	0.0270	21.2813	0.0470	259.9392	12.2145
29	1.7758	0.5631	38.7922	0.0258	21.8444	0.0458	275.7064	12.6214
30	1.8114	0.5521	40.5681	0.0246	22.3965	0.0446	291.7164	13.0251
36	2.0399	0.4902	51.9944	0.0192	25.4888	0.0392	392.0405	15.3809
42	2.2972	0.4353	64.8622	0.0154	28.2348	0.0354	497.6010	17.6237
48	2.5871	0.3865	79.3535	0.0126	30.6731	0.0326	605.9657	19.7556
54	2.9135	0.3432	95.6731	0.0105	32.8383	0.0305	715.1815	21.7789
60	3.2810	0.3048	114.0515	8.768E-03	34.7609	0.0288	823.6975	23.6961
66	3.6950	0.2706	134.7487	7.421E-03	36.4681	0.0274	930.3000	25.5100
72	4.1611	0.2403	158.0570	6.327E-03	37.9841	0.0263	1.034E+03	27.2234
120	10.7652	0.0929	488.2582	2.048E-03	45.3554	0.0220	1.710E+03	37.7114
180	35.3208	0.0283	1.716E+03	5.827E-04	48.5844	0.0206	2.174E+03	44.7554
360	1.248E+03	8.016E-04	6.233E+04	1.604E-05	49.9599	0.0200	2.484E+03	49.7112

Time Value of Money Factors—Discrete Compounding
i = 3%

	Single Sums		Uniform Series				Gradient Series	
	To Find F Given P	To Find P Given F	To Find F Given A	To Find A Given F	To Find P Given A	To Find A Given P	To Find P Given G	To Find A Given G
n	(F|P,i%,n)	(P|F,i%,n)	(F|A,i%,n)	(A|F,i%,n)	(P|A,i%,n)	(A|P,i%,n)	(P|G,i%,n)	(A|G,i%,n)
1	1.0300	0.9709	1.0000	1.0000	0.9709	1.0300	0.0000	0.0000
2	1.0609	0.9426	2.0300	0.4926	1.9135	0.5226	0.9426	0.4926
3	1.0927	0.9151	3.0909	0.3235	2.8286	0.3535	2.7729	0.9803
4	1.1255	0.8885	4.1836	0.2390	3.7171	0.2690	5.4383	1.4631
5	1.1593	0.8626	5.3091	0.1884	4.5797	0.2184	8.8888	1.9409
6	1.1941	0.8375	6.4684	0.1546	5.4172	0.1846	13.0762	2.4138
7	1.2299	0.8131	7.6625	0.1305	6.2303	0.1605	17.9547	2.8819
8	1.2668	0.7894	8.8923	0.1125	7.0197	0.1425	23.4806	3.3450
9	1.3048	0.7664	10.1591	0.0984	7.7861	0.1284	29.6119	3.8032
10	1.3439	0.7441	11.4639	0.0872	8.5302	0.1172	36.3088	4.2565
11	1.3842	0.7224	12.8078	0.0781	9.2526	0.1081	43.5330	4.7049
12	1.4258	0.7014	14.1920	0.0705	9.9540	0.1005	51.2482	5.1485
13	1.4685	0.6810	15.6178	0.0640	10.6350	0.0940	59.4196	5.5872
14	1.5126	0.6611	17.0863	0.0585	11.2961	0.0885	68.0141	6.0210
15	1.5580	0.6419	18.5989	0.0538	11.9379	0.0838	77.0002	6.4500
16	1.6047	0.6232	20.1569	0.0496	12.5611	0.0796	86.3477	6.8742
17	1.6528	0.6050	21.7616	0.0460	13.1661	0.0760	96.0280	7.2936
18	1.7024	0.5874	23.4144	0.0427	13.7535	0.0727	106.0137	7.7081
19	1.7535	0.5703	25.1169	0.0398	14.3238	0.0698	116.2788	8.1179
20	1.8061	0.5537	26.8704	0.0372	14.8775	0.0672	126.7987	8.5229
21	1.8603	0.5375	28.6765	0.0349	15.4150	0.0649	137.5496	8.9231
22	1.9161	0.5219	30.5368	0.0327	15.9369	0.0627	148.5094	9.3186
23	1.9736	0.5067	32.4529	0.0308	16.4436	0.0608	159.6566	9.7093
24	2.0328	0.4919	34.4265	0.0290	16.9355	0.0590	170.9711	10.0954
25	2.0938	0.4776	36.4593	0.0274	17.4131	0.0574	182.4336	10.4768
26	2.1566	0.4637	38.5530	0.0259	17.8768	0.0559	194.0260	10.8535
27	2.2213	0.4502	40.7096	0.0246	18.3270	0.0546	205.7309	11.2255
28	2.2879	0.4371	42.9309	0.0233	18.7641	0.0533	217.5320	11.5930
29	2.3566	0.4243	45.2189	0.0221	19.1885	0.0521	229.4137	11.9558
30	2.4273	0.4120	47.5754	0.0210	19.6004	0.0510	241.3613	12.3141
36	2.8983	0.3450	63.2759	0.0158	21.8323	0.0458	313.7028	14.3688
42	3.4607	0.2890	82.0232	0.0122	23.7014	0.0422	385.5024	16.2650
48	4.1323	0.2420	104.4084	9.578E-03	25.2667	0.0396	455.0255	18.0089
54	4.9341	0.2027	131.1375	7.626E-03	26.5777	0.0376	521.1157	19.6073
60	5.8916	0.1697	163.0534	6.133E-03	27.6756	0.0361	583.0526	21.0674
66	7.0349	0.1421	201.1627	4.971E-03	28.5950	0.0350	640.4407	22.3969
72	8.4000	0.1190	246.6672	4.054E-03	29.3651	0.0341	693.1226	23.6036
120	34.7110	0.0288	1.124E+03	8.899E-04	32.3730	0.0309	963.8635	29.7737
180	204.5034	4.890E-03	6.783E+03	1.474E-04	33.1703	0.0301	1.076E+03	32.4488
360	4.182E+04	2.391E-05	1.394E+06	7.173E-07	33.3325	0.0300	1.111E+03	33.3247

Time Value of Money Factors — Discrete Compounding
i = 4%

	Single Sums		Uniform Series				Gradient Series	
	To Find F Given P (F|P,i%,n)	To Find P Given F (P|F,i%,n)	To Find F Given A (F|A,i%,n)	To Find A Given F (A|F,i%,n)	To Find P Given A (P|A,i%,n)	To Find A Given P (A|P,i%,n)	To Find P Given G (P|G,i%,n)	To Find A Given G (A|G,i%,n)
n								
1	1.0400	0.9615	1.0000	1.0000	0.9615	1.0400	0.0000	0.0000
2	1.0816	0.9246	2.0400	0.4902	1.8861	0.5302	0.9246	0.4902
3	1.1249	0.8890	3.1216	0.3203	2.7751	0.3603	2.7025	0.9739
4	1.1699	0.8548	4.2465	0.2355	3.6299	0.2755	5.2670	1.4510
5	1.2167	0.8219	5.4163	0.1846	4.4518	0.2246	8.5547	1.9216
6	1.2653	0.7903	6.6330	0.1508	5.2421	0.1908	12.5062	2.3857
7	1.3159	0.7599	7.8983	0.1266	6.0021	0.1666	17.0657	2.8433
8	1.3686	0.7307	9.2142	0.1085	6.7327	0.1485	22.1806	3.2944
9	1.4233	0.7026	10.5828	0.0945	7.4353	0.1345	27.8013	3.7391
10	1.4802	0.6756	12.0061	0.0833	8.1109	0.1233	33.8814	4.1773
11	1.5395	0.6496	13.4864	0.0741	8.7605	0.1141	40.3772	4.6090
12	1.6010	0.6246	15.0258	0.0666	9.3851	0.1066	47.2477	5.0343
13	1.6651	0.6006	16.6268	0.0601	9.9856	0.1001	54.4546	5.4533
14	1.7317	0.5775	18.2919	0.0547	10.5631	0.0947	61.9618	5.8659
15	1.8009	0.5553	20.0236	0.0499	11.1184	0.0899	69.7355	6.2721
16	1.8730	0.5339	21.8245	0.0458	11.6523	0.0858	77.7441	6.6720
17	1.9479	0.5134	23.6975	0.0422	12.1657	0.0822	85.9581	7.0656
18	2.0258	0.4936	25.6454	0.0390	12.6593	0.0790	94.3498	7.4530
19	2.1068	0.4746	27.6712	0.0361	13.1339	0.0761	102.8933	7.8342
20	2.1911	0.4564	29.7781	0.0336	13.5903	0.0736	111.5647	8.2091
21	2.2788	0.4388	31.9692	0.0313	14.0292	0.0713	120.3414	8.5779
22	2.3699	0.4220	34.2480	0.0292	14.4511	0.0692	129.2024	8.9407
23	2.4647	0.4057	36.6179	0.0273	14.8568	0.0673	138.1284	9.2973
24	2.5633	0.3901	39.0826	0.0256	15.2470	0.0656	147.1012	9.6479
25	2.6658	0.3751	41.6459	0.0240	15.6221	0.0640	156.1040	9.9925
26	2.7725	0.3607	44.3117	0.0226	15.9828	0.0626	165.1212	10.3312
27	2.8834	0.3468	47.0842	0.0212	16.3296	0.0612	174.1385	10.6640
28	2.9987	0.3335	49.9676	0.0200	16.6631	0.0600	183.1424	10.9909
29	3.1187	0.3207	52.9663	0.0189	16.9837	0.0589	192.1206	11.3120
30	3.2434	0.3083	56.0849	0.0178	17.2920	0.0578	201.0618	11.6274
36	4.1039	0.2437	77.5983	0.0129	18.9083	0.0529	253.4052	13.4018
42	5.1928	0.1926	104.8196	9.540E-03	20.1856	0.0495	302.4370	14.9828
48	6.5705	0.1522	139.2632	7.181E-03	21.1951	0.0472	347.2446	16.3832
54	8.3138	0.1203	182.8454	5.469E-03	21.9930	0.0455	387.4436	17.6167
60	10.5196	0.0951	237.9907	4.202E-03	22.6235	0.0442	422.9966	18.6972
66	13.3107	0.0751	307.7671	3.249E-03	23.1218	0.0432	454.0847	19.6388
72	16.8423	0.0594	396.0566	2.525E-03	23.5156	0.0425	481.0170	20.4552
120	110.6626	9.036E-03	2.742E+03	3.648E-04	24.7741	0.0404	592.2428	23.9057
180	1.164E+03	8.590E-04	2.908E+04	3.439E-05	24.9785	0.0400	620.5976	24.8452
360	1.355E+06	7.379E-07	3.388E+07	2.952E-08	25.0000	0.0400	624.9929	24.9997

Time Value of Money Factors—Discrete Compounding
i = 5%

n	Single Sums		Uniform Series				Gradient Series	
	To Find F Given P (F|P,i%,n)	To Find P Given F (P|F,i%,n)	To Find F Given A (F|A,i%,n)	To Find A Given F (A|F,i%,n)	To Find P Given A (P|A,i%,n)	To Find A Given P (A|P,i%,n)	To Find P Given G (P|G,i%,n)	To Find A Given G (A|G,i%,n)
1	1.0500	0.9524	1.0000	1.0000	0.9524	1.0500	0.0000	0.0000
2	1.1025	0.9070	2.0500	0.4878	1.8594	0.5378	0.9070	0.4878
3	1.1576	0.8638	3.1525	0.3172	2.7232	0.3672	2.6347	0.9675
4	1.2155	0.8227	4.3101	0.2320	3.5460	0.2820	5.1028	1.4391
5	1.2763	0.7835	5.5256	0.1810	4.3295	0.2310	8.2369	1.9025
6	1.3401	0.7462	6.8019	0.1470	5.0757	0.1970	11.9680	2.3579
7	1.4071	0.7107	8.1420	0.1228	5.7864	0.1728	16.2321	2.8052
8	1.4775	0.6768	9.5491	0.1047	6.4632	0.1547	20.9700	3.2445
9	1.5513	0.6446	11.0266	0.0907	7.1078	0.1407	26.1268	3.6758
10	1.6289	0.6139	12.5779	0.0795	7.7217	0.1295	31.6520	4.0991
11	1.7103	0.5847	14.2068	0.0704	8.3064	0.1204	37.4988	4.5144
12	1.7959	0.5568	15.9171	0.0628	8.8633	0.1128	43.6241	4.9219
13	1.8856	0.5303	17.7130	0.0565	9.3936	0.1065	49.9879	5.3215
14	1.9799	0.5051	19.5986	0.0510	9.8986	0.1010	56.5538	5.7133
15	2.0789	0.4810	21.5786	0.0463	10.3797	0.0963	63.2880	6.0973
16	2.1829	0.4581	23.6575	0.0423	10.8378	0.0923	70.1597	6.4736
17	2.2920	0.4363	25.8404	0.0387	11.2741	0.0887	77.1405	6.8423
18	2.4066	0.4155	28.1324	0.0355	11.6896	0.0855	84.2043	7.2034
19	2.5270	0.3957	30.5390	0.0327	12.0853	0.0827	91.3275	7.5569
20	2.6533	0.3769	33.0660	0.0302	12.4622	0.0802	98.4884	7.9030
21	2.7860	0.3589	35.7193	0.0280	12.8212	0.0780	105.6673	8.2416
22	2.9253	0.3418	38.5052	0.0260	13.1630	0.0760	112.8461	8.5730
23	3.0715	0.3256	41.4305	0.0241	13.4886	0.0741	120.0087	8.8971
24	3.2251	0.3101	44.5020	0.0225	13.7986	0.0725	127.1402	9.2140
25	3.3864	0.2953	47.7271	0.0210	14.0939	0.0710	134.2275	9.5238
26	3.5557	0.2812	51.1135	0.0196	14.3752	0.0696	141.2585	9.8266
27	3.7335	0.2678	54.6691	0.0183	14.6430	0.0683	148.2226	10.1224
28	3.9201	0.2551	58.4026	0.0171	14.8981	0.0671	155.1101	10.4114
29	4.1161	0.2429	62.3227	0.0160	15.1411	0.0660	161.9126	10.6936
30	4.3219	0.2314	66.4388	0.0151	15.3725	0.0651	168.6226	10.9691
36	5.7918	0.1727	95.8363	0.0104	16.5469	0.0604	206.6237	12.4872
42	7.7616	0.1288	135.2318	7.395E-03	17.4232	0.0574	240.2389	13.7884
48	10.4013	0.0961	188.0254	5.318E-03	18.0772	0.0553	269.2467	14.8943
54	13.9387	0.0717	258.7739	3.864E-03	18.5651	0.0539	293.8208	15.8265
60	18.6792	0.0535	353.5837	2.828E-03	18.9293	0.0528	314.3432	16.6062
66	25.0319	0.0399	480.6379	2.081E-03	19.2010	0.0521	331.2877	17.2536
72	33.5451	0.0298	650.9027	1.536E-03	19.4038	0.0515	345.1485	17.7877
120	348.9120	2.866E-03	6.958E+03	1.437E-04	19.9427	0.0501	391.9751	19.6551
180	6.517E+03	1.534E-04	1.303E+05	7.673E-06	19.9969	0.0500	399.3863	19.9724
360	4.248E+07	2.354E-08	8.495E+08	1.177E-09	20.0000	0.0500	399.9998	20.0000

Time Value of Money Factors—Discrete Compounding
i = 6%

n	Single Sums		Uniform Series				Gradient Series	
	To Find F Given P (F|P,i%,n)	To Find P Given F (P|F,i%,n)	To Find F Given A (F|A,i%,n)	To Find A Given F (A|F,i%,n)	To Find P Given A (P|A,i%,n)	To Find A Given P (A|P,i%,n)	To Find P Given G (P|G,i%,n)	To Find A Given G (A|G,i%,n)
1	1.0600	0.9434	1.0000	1.0000	0.9434	1.0600	0.0000	0.0000
2	1.1236	0.8900	2.0600	0.4854	1.8334	0.5454	0.8900	0.4854
3	1.1910	0.8396	3.1836	0.3141	2.6730	0.3741	2.5692	0.9612
4	1.2625	0.7921	4.3746	0.2286	3.4651	0.2886	4.9455	1.4272
5	1.3382	0.7473	5.6371	0.1774	4.2124	0.2374	7.9345	1.8836
6	1.4185	0.7050	6.9753	0.1434	4.9173	0.2034	11.4594	2.3304
7	1.5036	0.6651	8.3938	0.1191	5.5824	0.1791	15.4497	2.7676
8	1.5938	0.6274	9.8975	0.1010	6.2098	0.1610	19.8416	3.1952
9	1.6895	0.5919	11.4913	0.0870	6.8017	0.1470	24.5768	3.6133
10	1.7908	0.5584	13.1808	0.0759	7.3601	0.1359	29.6023	4.0220
11	1.8983	0.5268	14.9716	0.0668	7.8869	0.1268	34.8702	4.4213
12	2.0122	0.4970	16.8699	0.0593	8.3838	0.1193	40.3369	4.8113
13	2.1329	0.4688	18.8821	0.0530	8.8527	0.1130	45.9629	5.1920
14	2.2609	0.4423	21.0151	0.0476	9.2950	0.1076	51.7128	5.5635
15	2.3966	0.4173	23.2760	0.0430	9.7122	0.1030	57.5546	5.9260
16	2.5404	0.3936	25.6725	0.0390	10.1059	0.0990	63.4592	6.2794
17	2.6928	0.3714	28.2129	0.0354	10.4773	0.0954	69.4011	6.6240
18	2.8543	0.3503	30.9057	0.0324	10.8276	0.0924	75.3569	6.9597
19	3.0256	0.3305	33.7600	0.0296	11.1581	0.0896	81.3062	7.2867
20	3.2071	0.3118	36.7856	0.0272	11.4699	0.0872	87.2304	7.6051
21	3.3996	0.2942	39.9927	0.0250	11.7641	0.0850	93.1136	7.9151
22	3.6035	0.2775	43.3923	0.0230	12.0416	0.0830	98.9412	8.2166
23	3.8197	0.2618	46.9958	0.0213	12.3034	0.0813	104.7007	8.5099
24	4.0489	0.2470	50.8156	0.0197	12.5504	0.0797	110.3812	8.7951
25	4.2919	0.2330	54.8645	0.0182	12.7834	0.0782	115.9732	9.0722
26	4.5494	0.2198	59.1564	0.0169	13.0032	0.0769	121.4684	9.3414
27	4.8223	0.2074	63.7058	0.0157	13.2105	0.0757	126.8600	9.6029
28	5.1117	0.1956	68.5281	0.0146	13.4062	0.0746	132.1420	9.8568
29	5.4184	0.1846	73.6398	0.0136	13.5907	0.0736	137.3096	10.1032
30	5.7435	0.1741	79.0582	0.0126	13.7648	0.0726	142.3588	10.3422
36	8.1473	0.1227	119.1209	8.395E-03	14.6210	0.0684	170.0387	11.6298
42	11.5570	0.0865	175.9505	5.683E-03	15.2245	0.0657	193.1732	12.6883
48	16.3939	0.0610	256.5645	3.898E-03	15.6500	0.0639	212.0351	13.5485
54	23.2550	0.0430	370.9170	2.696E-03	15.9500	0.0627	227.1316	14.2402
60	32.9877	0.0303	533.1282	1.876E-03	16.1614	0.0619	239.0428	14.7909
66	46.7937	0.0214	763.2278	1.310E-03	16.3105	0.0613	248.3341	15.2254
72	66.3777	0.0151	1.090E+03	9.177E-04	16.4156	0.0609	255.5146	15.5654
120	1.088E+03	9.190E-04	1.812E+04	5.519E-05	16.6514	0.0601	275.6846	16.5563
180	3.590E+04	2.786E-05	5.983E+05	1.672E-06	16.6662	0.0600	277.6865	16.6617
360	1.289E+09	7.760E-10	2.148E+10	4.656E-11	16.6667	0.0600	277.7778	16.6667

Time Value of Money Factors—Discrete Compounding
i = 7%

	Single Sums		Uniform Series				Gradient Series	
	To Find F Given P (F|P,i%,n)	To Find P Given F (P|F,i%,n)	To Find F Given A (F|A,i%,n)	To Find A Given F (A|F,i%,n)	To Find P Given A (P|A,i%,n)	To Find A Given P (A|P,i%,n)	To Find P Given G (P|G,i%,n)	To Find A Given G (A|G,i%,n)
n								
1	1.0700	0.9346	1.0000	1.0000	0.9346	1.0700	0.0000	0.0000
2	1.1449	0.8734	2.0700	0.4831	1.8080	0.5531	0.8734	0.4831
3	1.2250	0.8163	3.2149	0.3111	2.6243	0.3811	2.5060	0.9549
4	1.3108	0.7629	4.4399	0.2252	3.3872	0.2952	4.7947	1.4155
5	1.4026	0.7130	5.7507	0.1739	4.1002	0.2439	7.6467	1.8650
6	1.5007	0.6663	7.1533	0.1398	4.7665	0.2098	10.9784	2.3032
7	1.6058	0.6227	8.6540	0.1156	5.3893	0.1856	14.7149	2.7304
8	1.7182	0.5820	10.2598	0.0975	5.9713	0.1675	18.7889	3.1465
9	1.8385	0.5439	11.9780	0.0835	6.5152	0.1535	23.1404	3.5517
10	1.9672	0.5083	13.8164	0.0724	7.0236	0.1424	27.7156	3.9461
11	2.1049	0.4751	15.7836	0.0634	7.4987	0.1334	32.4665	4.3296
12	2.2522	0.4440	17.8885	0.0559	7.9427	0.1259	37.3506	4.7025
13	2.4098	0.4150	20.1406	0.0497	8.3577	0.1197	42.3302	5.0648
14	2.5785	0.3878	22.5505	0.0443	8.7455	0.1143	47.3718	5.4167
15	2.7590	0.3624	25.1290	0.0398	9.1079	0.1098	52.4461	5.7583
16	2.9522	0.3387	27.8881	0.0359	9.4466	0.1059	57.5271	6.0897
17	3.1588	0.3166	30.8402	0.0324	9.7632	0.1024	62.5923	6.4110
18	3.3799	0.2959	33.9990	0.0294	10.0591	0.0994	67.6219	6.7225
19	3.6165	0.2765	37.3790	0.0268	10.3356	0.0968	72.5991	7.0242
20	3.8697	0.2584	40.9955	0.0244	10.5940	0.0944	77.5091	7.3163
21	4.1406	0.2415	44.8652	0.0223	10.8355	0.0923	82.3393	7.5990
22	4.4304	0.2257	49.0057	0.0204	11.0612	0.0904	87.0793	7.8725
23	4.7405	0.2109	53.4361	0.0187	11.2722	0.0887	91.7201	8.1369
24	5.0724	0.1971	58.1767	0.0172	11.4693	0.0872	96.2545	8.3923
25	5.4274	0.1842	63.2490	0.0158	11.6536	0.0858	100.6765	8.6391
26	5.8074	0.1722	68.6765	0.0146	11.8258	0.0846	104.9814	8.8773
27	6.2139	0.1609	74.4838	0.0134	11.9867	0.0834	109.1656	9.1072
28	6.6488	0.1504	80.6977	0.0124	12.1371	0.0824	113.2264	9.3289
29	7.1143	0.1406	87.3465	0.0114	12.2777	0.0814	117.1622	9.5427
30	7.6123	0.1314	94.4608	0.0106	12.4090	0.0806	120.9718	9.7487
36	11.4239	0.0875	148.9135	6.715E-03	13.0352	0.0767	141.1990	10.8321
42	17.1443	0.0583	230.6322	4.336E-03	13.4524	0.0743	157.1807	11.6842
48	25.7289	0.0389	353.2701	2.831E-03	13.7305	0.0728	169.4981	12.3447
54	38.6122	0.0259	537.3164	1.861E-03	13.9157	0.0719	178.8173	12.8500
60	57.9464	0.0173	813.5204	1.229E-03	14.0392	0.0712	185.7677	13.2321
66	86.9620	0.0115	1.228E+03	8.143E-04	14.1214	0.0708	190.8927	13.5179
72	130.5065	7.662E-03	1.850E+03	5.405E-04	14.1763	0.0705	194.6365	13.7298
120	3.358E+03	2.978E-04	4.795E+04	2.085E-05	14.2815	0.0700	203.5103	14.2500
180	1.946E+05	5.139E-06	2.780E+06	3.598E-07	14.2856	0.0700	204.0674	14.2848
360	3.786E+10	2.641E-11	5.408E+11	1.849E-12	14.2857	0.0700	204.0816	14.2857

Time Value of Money Factors—Discrete Compounding
i = 8%

	Single Sums		Uniform Series				Gradient Series	
	To Find F Given P	To Find P Given F	To Find F Given A	To Find A Given F	To Find P Given A	To Find A Given P	To Find P Given G	To Find A Given G
n	(F|P,i%,n)	(P|F,i%,n)	(F|A,i%,n)	(A|F,i%,n)	(P|A,i%,n)	(A|P,i%,n)	(P|G,i%,n)	(A|G,i%,n)
1	1.0800	0.9259	1.0000	1.0000	0.9259	1.0800	0.0000	0.0000
2	1.1664	0.8573	2.0800	0.4808	1.7833	0.5608	0.8573	0.4808
3	1.2597	0.7938	3.2464	0.3080	2.5771	0.3880	2.4450	0.9487
4	1.3605	0.7350	4.5061	0.2219	3.3121	0.3019	4.6501	1.4040
5	1.4693	0.6806	5.8666	0.1705	3.9927	0.2505	7.3724	1.8465
6	1.5869	0.6302	7.3359	0.1363	4.6229	0.2163	10.5233	2.2763
7	1.7138	0.5835	8.9228	0.1121	5.2064	0.1921	14.0242	2.6937
8	1.8509	0.5403	10.6366	0.0940	5.7466	0.1740	17.8061	3.0985
9	1.9990	0.5002	12.4876	0.0801	6.2469	0.1601	21.8081	3.4910
10	2.1589	0.4632	14.4866	0.0690	6.7101	0.1490	25.9768	3.8713
11	2.3316	0.4289	16.6455	0.0601	7.1390	0.1401	30.2657	4.2395
12	2.5182	0.3971	18.9771	0.0527	7.5361	0.1327	34.6339	4.5957
13	2.7196	0.3677	21.4953	0.0465	7.9038	0.1265	39.0463	4.9402
14	2.9372	0.3405	24.2149	0.0413	8.2442	0.1213	43.4723	5.2731
15	3.1722	0.3152	27.1521	0.0368	8.5595	0.1168	47.8857	5.5945
16	3.4259	0.2919	30.3243	0.0330	8.8514	0.1130	52.2640	5.9046
17	3.7000	0.2703	33.7502	0.0296	9.1216	0.1096	56.5883	6.2037
18	3.9960	0.2502	37.4502	0.0267	9.3719	0.1067	60.8426	6.4920
19	4.3157	0.2317	41.4463	0.0241	9.6036	0.1041	65.0134	6.7697
20	4.6610	0.2145	45.7620	0.0219	9.8181	0.1019	69.0898	7.0369
21	5.0338	0.1987	50.4229	0.0198	10.0168	0.0998	73.0629	7.2940
22	5.4365	0.1839	55.4568	0.0180	10.2007	0.0980	76.9257	7.5412
23	5.8715	0.1703	60.8933	0.0164	10.3711	0.0964	80.6726	7.7786
24	6.3412	0.1577	66.7648	0.0150	10.5288	0.0950	84.2997	8.0066
25	6.8485	0.1460	73.1059	0.0137	10.6748	0.0937	87.8041	8.2254
26	7.3964	0.1352	79.9544	0.0125	10.8100	0.0925	91.1842	8.4352
27	7.9881	0.1252	87.3508	0.0114	10.9352	0.0914	94.4390	8.6363
28	8.6271	0.1159	95.3388	0.0105	11.0511	0.0905	97.5687	8.8289
29	9.3173	0.1073	103.9659	9.619E-03	11.1584	0.0896	100.5738	9.0133
30	10.0627	0.0994	113.2832	8.827E-03	11.2578	0.0888	103.4558	9.1897
36	15.9682	0.0626	187.1021	5.345E-03	11.7172	0.0853	118.2839	10.0949
42	25.3395	0.0395	304.2435	3.287E-03	12.0067	0.0833	129.3651	10.7744
48	40.2106	0.0249	490.1322	2.040E-03	12.1891	0.0820	137.4428	11.2758
54	63.8091	0.0157	785.1141	1.274E-03	12.3041	0.0813	143.2229	11.6403
60	101.2571	9.876E-03	1.253E+03	7.979E-04	12.3766	0.0808	147.3000	11.9015
66	160.6822	6.223E-03	1.996E+03	5.010E-04	12.4222	0.0805	150.1432	12.0867
72	254.9825	3.922E-03	3.175E+03	3.150E-04	12.4510	0.0803	152.1076	12.2165
120	1.025E+04	9.753E-05	1.281E+05	7.803E-06	12.4988	0.0800	156.0885	12.4883
180	1.038E+06	9.632E-07	1.298E+07	7.706E-08	12.5000	0.0800	156.2477	12.4998
360	1.078E+12	9.278E-13	1.347E+13	7.422E-14	12.5000	0.0800	156.2500	12.5000

Time Value of Money Factors—Discrete Compounding
i = 9%

	Single Sums		Uniform Series				Gradient Series	
	To Find F Given P (F|P,i%,n)	To Find P Given F (P|F,i%,n)	To Find F Given A (F|A,i%,n)	To Find A Given F (A|F,i%,n)	To Find P Given A (P|A,i%,n)	To Find A Given P (A|P,i%,n)	To Find P Given G (P|G,i%,n)	To Find A Given G (A|G,i%,n)
n								
1	1.0900	0.9174	1.0000	1.0000	0.9174	1.0900	0.0000	0.0000
2	1.1881	0.8417	2.0900	0.4785	1.7591	0.5685	0.8417	0.4785
3	1.2950	0.7722	3.2781	0.3051	2.5313	0.3951	2.3860	0.9426
4	1.4116	0.7084	4.5731	0.2187	3.2397	0.3087	4.5113	1.3925
5	1.5386	0.6499	5.9847	0.1671	3.8897	0.2571	7.1110	1.8282
6	1.6771	0.5963	7.5233	0.1329	4.4859	0.2229	10.0924	2.2498
7	1.8280	0.5470	9.2004	0.1087	5.0330	0.1987	13.3746	2.6574
8	1.9926	0.5019	11.0285	0.0907	5.5348	0.1807	16.8877	3.0512
9	2.1719	0.4604	13.0210	0.0768	5.9952	0.1668	20.5711	3.4312
10	2.3674	0.4224	15.1929	0.0658	6.4177	0.1558	24.3728	3.7978
11	2.5804	0.3875	17.5603	0.0569	6.8052	0.1469	28.2481	4.1510
12	2.8127	0.3555	20.1407	0.0497	7.1607	0.1397	32.1590	4.4910
13	3.0658	0.3262	22.9534	0.0436	7.4869	0.1336	36.0731	4.8182
14	3.3417	0.2992	26.0192	0.0384	7.7862	0.1284	39.9633	5.1326
15	3.6425	0.2745	29.3609	0.0341	8.0607	0.1241	43.8069	5.4346
16	3.9703	0.2519	33.0034	0.0303	8.3126	0.1203	47.5849	5.7245
17	4.3276	0.2311	36.9737	0.0270	8.5436	0.1170	51.2821	6.0024
18	4.7171	0.2120	41.3013	0.0242	8.7556	0.1142	54.8860	6.2687
19	5.1417	0.1945	46.0185	0.0217	8.9501	0.1117	58.3868	6.5236
20	5.6044	0.1784	51.1601	0.0195	9.1285	0.1095	61.7770	6.7674
21	6.1088	0.1637	56.7645	0.0176	9.2922	0.1076	65.0509	7.0006
22	6.6586	0.1502	62.8733	0.0159	9.4424	0.1059	68.2048	7.2232
23	7.2579	0.1378	69.5319	0.0144	9.5802	0.1044	71.2359	7.4357
24	7.9111	0.1264	76.7898	0.0130	9.7066	0.1030	74.1433	7.6384
25	8.6231	0.1160	84.7009	0.0118	9.8226	0.1018	76.9265	7.8316
26	9.3992	0.1064	93.3240	0.0107	9.9290	0.1007	79.5863	8.0156
27	10.2451	0.0976	102.7231	9.735E-03	10.0266	0.0997	82.1241	8.1906
28	11.1671	0.0895	112.9682	8.852E-03	10.1161	0.0989	84.5419	8.3571
29	12.1722	0.0822	124.1354	8.056E-03	10.1983	0.0981	86.8422	8.5154
30	13.2677	0.0754	136.3075	7.336E-03	10.2737	0.0973	89.0280	8.6657
36	22.2512	0.0449	236.1247	4.235E-03	10.6118	0.0942	99.9319	9.4171
42	37.3175	0.0268	403.5281	2.478E-03	10.8134	0.0925	107.6432	9.9546
48	62.5852	0.0160	684.2804	1.461E-03	10.9336	0.0915	112.9625	10.3317
54	104.9617	9.527E-03	1.155E+03	8.657E-04	11.0053	0.0909	116.5642	10.5917
60	176.0313	5.681E-03	1.945E+03	5.142E-04	11.0480	0.0905	118.9683	10.7683
66	295.2221	3.387E-03	3.269E+03	3.059E-04	11.0735	0.0903	120.5546	10.8868
72	495.1170	2.020E-03	5.490E+03	1.821E-04	11.0887	0.0902	121.5917	10.9654
120	3.099E+04	3.227E-05	3.443E+05	2.905E-06	11.1108	0.0900	123.4098	11.1072
180	5.455E+06	1.833E-07	6.061E+07	1.650E-08	11.1111	0.0900	123.4564	11.1111
360	2.975E+13	3.361E-14	3.306E+14	3.025E-15	11.1111	0.0900	123.4568	11.1111

Time Value of Money Factors—Discrete Compounding
i = 10%

n	Single Sums		Uniform Series				Gradient Series	
	To Find F Given P (F|P,i%,n)	To Find P Given F (P|F,i%,n)	To Find F Given A (F|A,i%,n)	To Find A Given F (A|F,i%,n)	To Find P Given A (P|A,i%,n)	To Find A Given P (A|P,i%,n)	To Find P Given G (P|G,i%,n)	To Find A Given G (A|G,i%,n)
1	1.1000	0.9091	1.0000	1.0000	0.9091	1.1000	0.0000	0.0000
2	1.2100	0.8264	2.1000	0.4762	1.7355	0.5762	0.8264	0.4762
3	1.3310	0.7513	3.3100	0.3021	2.4869	0.4021	2.3291	0.9366
4	1.4641	0.6830	4.6410	0.2155	3.1699	0.3155	4.3781	1.3812
5	1.6105	0.6209	6.1051	0.1638	3.7908	0.2638	6.8618	1.8101
6	1.7716	0.5645	7.7156	0.1296	4.3553	0.2296	9.6842	2.2236
7	1.9487	0.5132	9.4872	0.1054	4.8684	0.2054	12.7631	2.6216
8	2.1436	0.4665	11.4359	0.0874	5.3349	0.1874	16.0287	3.0045
9	2.3579	0.4241	13.5795	0.0736	5.7590	0.1736	19.4215	3.3724
10	2.5937	0.3855	15.9374	0.0627	6.1446	0.1627	22.8913	3.7255
11	2.8531	0.3505	18.5312	0.0540	6.4951	0.1540	26.3963	4.0641
12	3.1384	0.3186	21.3843	0.0468	6.8137	0.1468	29.9012	4.3884
13	3.4523	0.2897	24.5227	0.0408	7.1034	0.1408	33.3772	4.6988
14	3.7975	0.2633	27.9750	0.0357	7.3667	0.1357	36.8005	4.9955
15	4.1772	0.2394	31.7725	0.0315	7.6061	0.1315	40.1520	5.2789
16	4.5950	0.2176	35.9497	0.0278	7.8237	0.1278	43.4164	5.5493
17	5.0545	0.1978	40.5447	0.0247	8.0216	0.1247	46.5819	5.8071
18	5.5599	0.1799	45.5992	0.0219	8.2014	0.1219	49.6395	6.0526
19	6.1159	0.1635	51.1591	0.0195	8.3649	0.1195	52.5827	6.2861
20	6.7275	0.1486	57.2750	0.0175	8.5136	0.1175	55.4069	6.5081
21	7.4002	0.1351	64.0025	0.0156	8.6487	0.1156	58.1095	6.7189
22	8.1403	0.1228	71.4027	0.0140	8.7715	0.1140	60.6893	6.9189
23	8.9543	0.1117	79.5430	0.0126	8.8832	0.1126	63.1462	7.1085
24	9.8497	0.1015	88.4973	0.0113	8.9847	0.1113	65.4813	7.2881
25	10.8347	0.0923	98.3471	0.0102	9.0770	0.1102	67.6964	7.4580
26	11.9182	0.0839	109.1818	9.159E-03	9.1609	0.1092	69.7940	7.6186
27	13.1100	0.0763	121.0999	8.258E-03	9.2372	0.1083	71.7773	7.7704
28	14.4210	0.0693	134.2099	7.451E-03	9.3066	0.1075	73.6495	7.9137
29	15.8631	0.0630	148.6309	6.728E-03	9.3696	0.1067	75.4146	8.0489
30	17.4494	0.0573	164.4940	6.079E-03	9.4269	0.1061	77.0766	8.1762
36	30.9127	0.0323	299.1268	3.343E-03	9.6765	0.1033	85.1194	8.7965
42	54.7637	0.0183	537.6370	1.860E-03	9.8174	0.1019	90.5047	9.2188
48	97.0172	0.0103	960.1723	1.041E-03	9.8969	0.1010	94.0217	9.5001
54	171.8719	5.818E-03	1.709E+03	5.852E-04	9.9418	0.1006	96.2763	9.6840
60	304.4816	3.284E-03	3.035E+03	3.295E-04	9.9672	0.1003	97.7010	9.8023
66	539.4078	1.854E-03	5.384E+03	1.857E-04	9.9815	0.1002	98.5910	9.8774
72	955.5938	1.046E-03	9.546E+03	1.048E-04	9.9895	0.1001	99.1419	9.9246
120	9.271E+04	1.079E-05	9.271E+05	1.079E-06	9.9999	0.1000	99.9860	9.9987
180	2.823E+07	3.543E-08	2.823E+08	3.543E-09	10.0000	0.1000	99.9999	10.0000
360	7.968E+14	1.255E-15	7.968E+15	1.255E-16	10.0000	0.1000	100.0000	10.0000

Time Value of Money Factors—Discrete Compounding
i = 12%

	Single Sums		Uniform Series				Gradient Series	
	To Find F Given P (F|P,i%,n)	To Find P Given F (P|F,i%,n)	To Find F Given A (F|A,i%,n)	To Find A Given F (A|F,i%,n)	To Find P Given A (P|A,i%,n)	To Find A Given P (A|P,i%,n)	To Find P Given G (P|G,i%,n)	To Find A Given G (A|G,i%,n)
n								
1	1.1200	0.8929	1.0000	1.0000	0.8929	1.1200	0.0000	0.0000
2	1.2544	0.7972	2.1200	0.4717	1.6901	0.5917	0.7972	0.4717
3	1.4049	0.7118	3.3744	0.2963	2.4018	0.4163	2.2208	0.9246
4	1.5735	0.6355	4.7793	0.2092	3.0373	0.3292	4.1273	1.3589
5	1.7623	0.5674	6.3528	0.1574	3.6048	0.2774	6.3970	1.7746
6	1.9738	0.5066	8.1152	0.1232	4.1114	0.2432	8.9302	2.1720
7	2.2107	0.4523	10.0890	0.0991	4.5638	0.2191	11.6443	2.5515
8	2.4760	0.4039	12.2997	0.0813	4.9676	0.2013	14.4714	2.9131
9	2.7731	0.3606	14.7757	0.0677	5.3282	0.1877	17.3563	3.2574
10	3.1058	0.3220	17.5487	0.0570	5.6502	0.1770	20.2541	3.5847
11	3.4785	0.2875	20.6546	0.0484	5.9377	0.1684	23.1288	3.8953
12	3.8960	0.2567	24.1331	0.0414	6.1944	0.1614	25.9523	4.1897
13	4.3635	0.2292	28.0291	0.0357	6.4235	0.1557	28.7024	4.4683
14	4.8871	0.2046	32.3926	0.0309	6.6282	0.1509	31.3624	4.7317
15	5.4736	0.1827	37.2797	0.0268	6.8109	0.1468	33.9202	4.9803
16	6.1304	0.1631	42.7533	0.0234	6.9740	0.1434	36.3670	5.2147
17	6.8660	0.1456	48.8837	0.0205	7.1196	0.1405	38.6973	5.4353
18	7.6900	0.1300	55.7497	0.0179	7.2497	0.1379	40.9080	5.6427
19	8.6128	0.1161	63.4397	0.0158	7.3658	0.1358	42.9979	5.8375
20	9.6463	0.1037	72.0524	0.0139	7.4694	0.1339	44.9676	6.0202
21	10.8038	0.0926	81.6987	0.0122	7.5620	0.1322	46.8188	6.1913
22	12.1003	0.0826	92.5026	0.0108	7.6446	0.1308	48.5543	6.3514
23	13.5523	0.0738	104.6029	9.560E-03	7.7184	0.1296	50.1776	6.5010
24	15.1786	0.0659	118.1552	8.463E-03	7.7843	0.1285	51.6929	6.6406
25	17.0001	0.0588	133.3339	7.500E-03	7.8431	0.1275	53.1046	6.7708
26	19.0401	0.0525	150.3339	6.652E-03	7.8957	0.1267	54.4177	6.8921
27	21.3249	0.0469	169.3740	5.904E-03	7.9426	0.1259	55.6369	7.0049
28	23.8839	0.0419	190.6989	5.244E-03	7.9844	0.1252	56.7674	7.1098
29	26.7499	0.0374	214.5828	4.660E-03	8.0218	0.1247	57.8141	7.2071
30	29.9599	0.0334	241.3327	4.144E-03	8.0552	0.1241	58.7821	7.2974
36	59.1356	0.0169	484.4631	2.064E-03	8.1924	0.1221	63.1970	7.7141
42	116.7231	8.567E-03	964.3595	1.037E-03	8.2619	0.1210	65.8509	7.9704
48	230.3908	4.340E-03	1.912E+03	5.231E-04	8.2972	0.1205	67.4068	8.1241
54	454.7505	2.199E-03	3.781E+03	2.645E-04	8.3150	0.1203	68.3022	8.2143
60	897.5969	1.114E-03	7.472E+03	1.338E-04	8.3240	0.1201	68.8100	8.2664
66	1.772E+03	5.644E-04	1.476E+04	6.777E-05	8.3286	0.1201	69.0948	8.2961
72	3.497E+03	2.860E-04	2.913E+04	3.432E-05	8.3310	0.1200	69.2530	8.3127
120	8.057E+05	1.241E-06	6.714E+06	1.489E-07	8.3333	0.1200	69.4431	8.3332
180	7.232E+08	1.383E-09	6.026E+09	1.659E-10	8.3333	0.1200	69.4444	8.3333
360	5.230E+17	1.912E-18	4.358E+18	2.295E-19	8.3333	0.1200	69.4444	8.3333

Time Value of Money Factors—Discrete Compounding
i = 15%

	Single Sums		Uniform Series				Gradient Series	
	To Find F Given P (F|P,i%,n)	To Find P Given F (P|F,i%,n)	To Find F Given A (F|A,i%,n)	To Find A Given F (A|F,i%,n)	To Find P Given A (P|A,i%,n)	To Find A Given P (A|P,i%,n)	To Find P Given G (P|G,i%,n)	To Find A Given G (A|G,i%,n)
n								
1	1.1500	0.8696	1.0000	1.0000	0.8696	1.1500	0.0000	0.0000
2	1.3225	0.7561	2.1500	0.4651	1.6257	0.6151	0.7561	0.4651
3	1.5209	0.6575	3.4725	0.2880	2.2832	0.4380	2.0712	0.9071
4	1.7490	0.5718	4.9934	0.2003	2.8550	0.3503	3.7864	1.3263
5	2.0114	0.4972	6.7424	0.1483	3.3522	0.2983	5.7751	1.7228
6	2.3131	0.4323	8.7537	0.1142	3.7845	0.2642	7.9368	2.0972
7	2.6600	0.3759	11.0668	0.0904	4.1604	0.2404	10.1924	2.4498
8	3.0590	0.3269	13.7268	0.0729	4.4873	0.2229	12.4807	2.7813
9	3.5179	0.2843	16.7858	0.0596	4.7716	0.2096	14.7548	3.0922
10	4.0456	0.2472	20.3037	0.0493	5.0188	0.1993	16.9795	3.3832
11	4.6524	0.2149	24.3493	0.0411	5.2337	0.1911	19.1289	3.6549
12	5.3503	0.1869	29.0017	0.0345	5.4206	0.1845	21.1849	3.9082
13	6.1528	0.1625	34.3519	0.0291	5.5831	0.1791	23.1352	4.1438
14	7.0757	0.1413	40.5047	0.0247	5.7245	0.1747	24.9725	4.3624
15	8.1371	0.1229	47.5804	0.0210	5.8474	0.1710	26.6930	4.5650
16	9.3576	0.1069	55.7175	0.0179	5.9542	0.1679	28.2960	4.7522
17	10.7613	0.0929	65.0751	0.0154	6.0472	0.1654	29.7828	4.9251
18	12.3755	0.0808	75.8364	0.0132	6.1280	0.1632	31.1565	5.0843
19	14.2318	0.0703	88.2118	0.0113	6.1982	0.1613	32.4213	5.2307
20	16.3665	0.0611	102.4436	9.761E-03	6.2593	0.1598	33.5822	5.3651
21	18.8215	0.0531	118.8101	8.417E-03	6.3125	0.1584	34.6448	5.4883
22	21.6447	0.0462	137.6316	7.266E-03	6.3587	0.1573	35.6150	5.6010
23	24.8915	0.0402	159.2764	6.278E-03	6.3988	0.1563	36.4988	5.7040
24	28.6252	0.0349	184.1678	5.430E-03	0.4338	0.1554	37.3023	5.7979
25	32.9190	0.0304	212.7930	4.699E-03	6.4641	0.1547	38.0314	5.8834
26	37.8568	0.0264	245.7120	4.070E-03	6.4906	0.1541	38.6918	5.9612
27	43.5353	0.0230	283.5688	3.526E-03	6.5135	0.1535	39.2890	6.0319
28	50.0656	0.0200	327.1041	3.057E-03	6.5335	0.1531	39.8283	6.0960
29	57.5755	0.0174	377.1697	2.651E-03	6.5509	0.1527	40.3146	6.1541
30	66.2118	0.0151	434.7451	2.300E-03	6.5660	0.1523	40.7526	6.2066
36	153.1519	6.529E-03	1.014E+03	9.859E-04	6.6231	0.1510	42.5872	6.4301
42	354.2495	2.823E-03	2.355E+03	4.246E-04	6.6478	0.1504	43.5286	6.5478
48	819.4007	1.220E-03	5.456E+03	1.833E-04	6.6585	0.1502	43.9997	6.6080
54	1.895E+03	5.276E-04	1.263E+04	7.918E-05	6.6631	0.1501	44.2311	6.6382
60	4.384E+03	2.281E-04	2.922E+04	3.422E-05	6.6651	0.1500	44.3431	6.6530
66	1.014E+04	9.861E-05	6.760E+04	1.479E-05	6.6660	0.1500	44.3967	6.6602
72	2.346E+04	4.263E-05	1.564E+05	6.395E-06	6.6664	0.1500	44.4221	6.6636
120	1.922E+07	5.203E-08	1.281E+08	7.805E-09	6.6667	0.1500	44.4444	6.6667
180	8.426E+10	1.187E-11	5.617E+11	1.780E-12	6.6667	0.1500	44.4444	6.6667
360	7.099E+21	1.409E-22	4.733E+22	2.113E-23	6.6667	0.1500	44.4444	6.6667

Time Value of Money Factors—Discrete Compounding
i = 18%

	Single Sums		Uniform Series				Gradient Series	
	To Find F Given P	To Find P Given F	To Find F Given A	To Find A Given F	To Find P Given A	To Find A Given P	To Find P Given G	To Find A Given G
n	(F|P,i%,n)	(P|F,i%,n)	(F|A,i%,n)	(A|F,i%,n)	(P|A,i%,n)	(A|P,i%,n)	(P|G,i%,n)	(A|G,i%,n)
1	1.1800	0.8475	1.0000	1.0000	0.8475	1.1800	0.0000	0.0000
2	1.3924	0.7182	2.1800	0.4587	1.5656	0.6387	0.7182	0.4587
3	1.6430	0.6086	3.5724	0.2799	2.1743	0.4599	1.9354	0.8902
4	1.9388	0.5158	5.2154	0.1917	2.6901	0.3717	3.4828	1.2947
5	2.2878	0.4371	7.1542	0.1398	3.1272	0.3198	5.2312	1.6728
6	2.6996	0.3704	9.4420	0.1059	3.4976	0.2859	7.0834	2.0252
7	3.1855	0.3139	12.1415	0.0824	3.8115	0.2624	8.9670	2.3526
8	3.7589	0.2660	15.3270	0.0652	4.0776	0.2452	10.8292	2.6558
9	4.4355	0.2255	19.0859	0.0524	4.3030	0.2324	12.6329	2.9358
10	5.2338	0.1911	23.5213	0.0425	4.4941	0.2225	14.3525	3.1936
11	6.1759	0.1619	28.7551	0.0348	4.6560	0.2148	15.9716	3.4303
12	7.2876	0.1372	34.9311	0.0286	4.7932	0.2086	17.4811	3.6470
13	8.5994	0.1163	42.2187	0.0237	4.9095	0.2037	18.8765	3.8449
14	10.1472	0.0985	50.8180	0.0197	5.0081	0.1997	20.1576	4.0250
15	11.9737	0.0835	60.9653	0.0164	5.0916	0.1964	21.3269	4.1887
16	14.1290	0.0708	72.9390	0.0137	5.1624	0.1937	22.3885	4.3369
17	16.6722	0.0600	87.0680	0.0115	5.2223	0.1915	23.3482	4.4708
18	19.6733	0.0508	103.7403	9.639E-03	5.2732	0.1896	24.2123	4.5916
19	23.2144	0.0431	123.4135	8.103E-03	5.3162	0.1881	24.9877	4.7003
20	27.3930	0.0365	146.6280	6.820E-03	5.3527	0.1868	25.6813	4.7978
21	32.3238	0.0309	174.0210	5.746E-03	5.3837	0.1857	26.3000	4.8851
22	38.1421	0.0262	206.3448	4.846E-03	5.4099	0.1848	26.8506	4.9632
23	45.0076	0.0222	244.4868	4.090E-03	5.4321	0.1841	27.3394	5.0329
24	53.1090	0.0188	289.4945	3.454E-03	5.4509	0.1835	27.7725	5.0950
25	62.6686	0.0160	342.6035	2.919E-03	5.4669	0.1829	28.1555	5.1502
26	73.9490	0.0135	405.2721	2.467E-03	5.4804	0.1825	28.4935	5.1991
27	87.2598	0.0115	479.2211	2.087E-03	5.4919	0.1821	28.7915	5.2425
28	102.9666	9.712E-03	566.4809	1.765E-03	5.5016	0.1818	29.0537	5.2810
29	121.5005	8.230E-03	669.4475	1.494E-03	5.5098	0.1815	29.2842	5.3149
30	143.3706	6.975E-03	790.9480	1.264E-03	5.5168	0.1813	29.4864	5.3448
36	387.0368	2.584E-03	2.145E+03	4.663E-04	5.5412	0.1805	30.2677	5.4623
42	1.045E+03	9.571E-04	5.799E+03	1.724E-04	5.5502	0.1802	30.6113	5.5153
48	2.821E+03	3.545E-04	1.566E+04	6.384E-05	5.5536	0.1801	30.7587	5.5385
54	7.614E+03	1.313E-04	4.230E+04	2.364E-05	5.5548	0.1800	30.8207	5.5485
60	2.056E+04	4.865E-05	1.142E+05	8.757E-06	5.5553	0.1800	30.8465	5.5526
66	5.549E+04	1.802E-05	3.083E+05	3.244E-06	5.5555	0.1800	30.8570	5.5544
72	1.498E+05	6.676E-06	8.322E+05	1.202E-06	5.5555	0.1800	30.8613	5.5551
120	4.225E+08	2.367E-09	2.347E+09	4.260E-10	5.5556	0.1800	30.8642	5.5556
180	8.685E+12	1.151E-13	4.825E+13	2.073E-14	5.5556	0.1800	30.8642	5.5556
360	7.543E+25	1.326E-26	4.190E+26	2.386E-27	5.5556	0.1800	30.8642	5.5556

Time Value of Money Factors—Discrete Compounding
i = 20%

n	Single Sums		Uniform Series				Gradient Series	
	To Find F Given P (F|P,i%,n)	To Find P Given F (P|F,i%,n)	To Find F Given A (F|A,i%,n)	To Find A Given F (A|F,i%,n)	To Find P Given A (P|A,i%,n)	To Find A Given P (A|P,i%,n)	To Find P Given G (P|G,i%,n)	To Find A Given G (A|G,i%,n)
1	1.2000	0.8333	1.0000	1.0000	0.8333	1.2000	0.0000	0.0000
2	1.4400	0.6944	2.2000	0.4545	1.5278	0.6545	0.6944	0.4545
3	1.7280	0.5787	3.6400	0.2747	2.1065	0.4747	1.8519	0.8791
4	2.0736	0.4823	5.3680	0.1863	2.5887	0.3863	3.2986	1.2742
5	2.4883	0.4019	7.4416	0.1344	2.9906	0.3344	4.9061	1.6405
6	2.9860	0.3349	9.9299	0.1007	3.3255	0.3007	6.5806	1.9788
7	3.5832	0.2791	12.9159	0.0774	3.6046	0.2774	8.2551	2.2902
8	4.2998	0.2326	16.4991	0.0606	3.8372	0.2606	9.8831	2.5756
9	5.1598	0.1938	20.7989	0.0481	4.0310	0.2481	11.4335	2.8364
10	6.1917	0.1615	25.9587	0.0385	4.1925	0.2385	12.8871	3.0739
11	7.4301	0.1346	32.1504	0.0311	4.3271	0.2311	14.2330	3.2893
12	8.9161	0.1122	39.5805	0.0253	4.4392	0.2253	15.4667	3.4841
13	10.6993	0.0935	48.4966	0.0206	4.5327	0.2206	16.5883	3.6597
14	12.8392	0.0779	59.1959	0.0169	4.6106	0.2169	17.6008	3.8175
15	15.4070	0.0649	72.0351	0.0139	4.6755	0.2139	18.5095	3.9588
16	18.4884	0.0541	87.4421	0.0114	4.7296	0.2114	19.3208	4.0851
17	22.1861	0.0451	105.9306	9.440E-03	4.7746	0.2094	20.0419	4.1976
18	26.6233	0.0376	128.1167	7.805E-03	4.8122	0.2078	20.6805	4.2975
19	31.9480	0.0313	154.7400	6.462E-03	4.8435	0.2065	21.2439	4.3861
20	38.3376	0.0261	186.6880	5.357E-03	4.8696	0.2054	21.7395	4.4643
21	46.0051	0.0217	225.0256	4.444E-03	4.8913	0.2044	22.1742	4.5334
22	55.2061	0.0181	271.0307	3.690E-03	4.9094	0.2037	22.5546	4.5941
23	66.2474	0.0151	326.2369	3.065E-03	4.9245	0.2031	22.8867	4.6475
24	79.4968	0.0126	392.4842	2.548E-03	4.9371	0.2025	23.1760	4.6943
25	95.3962	0.0105	471.9811	2.119E-03	4.9476	0.2021	23.4276	4.7352
26	114.4755	8.735E-03	567.3773	1.762E-03	4.9563	0.2018	23.6460	4.7709
27	137.3706	7.280E-03	681.8528	1.467E-03	4.9636	0.2015	23.8353	4.8020
28	164.8447	6.066E-03	819.2233	1.221E-03	4.9697	0.2012	23.9991	4.8291
29	197.8136	5.055E-03	984.0680	1.016E-03	4.9747	0.2010	24.1406	4.8527
30	237.3763	4.213E-03	1.182E+03	8.461E-04	4.9789	0.2008	24.2628	4.8731
36	708.8019	1.411E-03	3.539E+03	2.826E-04	4.9929	0.2003	24.7108	4.9491
42	2.116E+03	4.725E-04	1.058E+04	9.454E-05	4.9976	0.2001	24.8890	4.9801
48	6.320E+03	1.582E-04	3.159E+04	3.165E-05	4.9992	0.2000	24.9581	4.9924
54	1.887E+04	5.299E-05	9.435E+04	1.060E-05	4.9997	0.2000	24.9844	4.9971
60	5.635E+04	1.775E-05	2.817E+05	3.549E-06	4.9999	0.2000	24.9942	4.9989
66	1.683E+05	5.943E-06	8.413E+05	1.189E-06	5.0000	0.2000	24.9979	4.9996
72	5.024E+05	1.990E-06	2.512E+06	3.981E-07	5.0000	0.2000	24.9992	4.9999
120	3.175E+09	3.150E-10	1.588E+10	6.299E-11	5.0000	0.2000	25.0000	5.0000
180	1.789E+14	5.590E-15	8.945E+14	1.118E-15	5.0000	0.2000	25.0000	5.0000
360	3.201E+28	3.124E-29	1.600E+29	6.249E-30	5.0000	0.2000	25.0000	5.0000

Time Value of Money Factors—Discrete Compounding
i = 25%

	Single Sums		Uniform Series				Gradient Series	
	To Find F Given P	To Find P Given F	To Find F Given A	To Find A Given F	To Find P Given A	To Find A Given P	To Find P Given G	To Find A Given G
n	(F|P,i%,n)	(P|F,i%,n)	(F|A,i%,n)	(A|F,i%,n)	(P|A,i%,n)	(A|P,i%,n)	(P|G,i%,n)	(A|G,i%,n)
1	1.2500	0.8000	1.0000	1.0000	0.8000	1.2500	0.0000	0.0000
2	1.5625	0.6400	2.2500	0.4444	1.4400	0.6944	0.6400	0.4444
3	1.9531	0.5120	3.8125	0.2623	1.9520	0.5123	1.6640	0.8525
4	2.4414	0.4096	5.7656	0.1734	2.3616	0.4234	2.8928	1.2249
5	3.0518	0.3277	8.2070	0.1218	2.6893	0.3718	4.2035	1.5631
6	3.8147	0.2621	11.2588	0.0888	2.9514	0.3388	5.5142	1.8683
7	4.7684	0.2097	15.0735	0.0663	3.1611	0.3163	6.7725	2.1424
8	5.9605	0.1678	19.8419	0.0504	3.3289	0.3004	7.9469	2.3872
9	7.4506	0.1342	25.8023	0.0388	3.4631	0.2888	9.0207	2.6048
10	9.3132	0.1074	33.2529	0.0301	3.5705	0.2801	9.9870	2.7971
11	11.6415	0.0859	42.5661	0.0235	3.6564	0.2735	10.8460	2.9663
12	14.5519	0.0687	54.2077	0.0184	3.7251	0.2684	11.6020	3.1145
13	18.1899	0.0550	68.7596	0.0145	3.7801	0.2645	12.2617	3.2437
14	22.7374	0.0440	86.9495	0.0115	3.8241	0.2615	12.8334	3.3559
15	28.4217	0.0352	109.8668	9.117E-03	3.8593	0.2591	13.3260	3.4530
16	35.5271	0.0281	138.1085	7.241E-03	3.8874	0.2572	13.7482	3.5366
17	44.4089	0.0225	173.6357	5.759E-03	3.9099	0.2558	14.1085	3.6084
18	55.5112	0.0180	218.0446	4.586E-03	3.9279	0.2546	14.4147	3.6698
19	69.3889	0.0144	273.5558	3.656E-03	3.9424	0.2537	14.6741	3.7222
20	86.7362	0.0115	342.9447	2.916E-03	3.9539	0.2529	14.8932	3.7667
21	108.4202	9.223E-03	429.6809	2.327E-03	3.9631	0.2523	15.0777	3.8045
22	135.5253	7.379E-03	538.1011	1.858E-03	3.9705	0.2519	15.2326	3.8365
23	169.4066	5.903E-03	673.6264	1.485E-03	3.9764	0.2515	15.3625	3.8634
24	211.7582	4.722E-03	843.0329	1.186E-03	3.9811	0.2512	15.4711	3.8861
25	264.6978	3.778E-03	1.055E+03	9.481E-04	3.9849	0.2509	15.5618	3.9052
26	330.8722	3.022E-03	1.319E+03	7.579E-04	3.9879	0.2508	15.6373	3.9212
27	413.5903	2.418E-03	1.650E+03	6.059E-04	3.9903	0.2506	15.7002	3.9346
28	516.9879	1.934E-03	2.064E+03	4.845E-04	3.9923	0.2505	15.7524	3.9457
29	646.2349	1.547E-03	2.581E+03	3.875E-04	3.9938	0.2504	15.7957	3.9551
30	807.7936	1.238E-03	3.227E+03	3.099E-04	3.9950	0.2503	15.8316	3.9628
36	3.081E+03	3.245E-04	1.232E+04	8.116E-05	3.9987	0.2501	15.9481	3.9883
42	1.175E+04	8.507E-05	4.702E+04	2.127E-05	3.9997	0.2500	15.9843	3.9964
48	4.484E+04	2.230E-05	1.794E+05	5.575E-06	3.9999	0.2500	15.9954	3.9989
54	1.711E+05	5.846E-06	6.842E+05	1.462E-06	4.0000	0.2500	15.9986	3.9997
60	6.525E+05	1.532E-06	2.610E+06	3.831E-07	4.0000	0.2500	15.9996	3.9999
66	2.489E+06	4.017E-07	9.957E+06	1.004E-07	4.0000	0.2500	15.9999	4.0000
72	9.496E+06	1.053E-07	3.798E+07	2.633E-08	4.0000	0.2500	16.0000	4.0000
120	4.258E+11	2.349E-12	1.703E+12	5.871E-13	4.0000	0.2500	16.0000	4.0000
180	2.778E+17	3.599E-18	1.111E+18	8.998E-19	4.0000	0.2500	16.0000	4.0000
360	7.720E+34	1.295E-35	3.088E+35	3.238E-36	4.0000	0.2500	16.0000	4.0000

Time Value of Money Factors—Discrete Compounding
i = 30%

	Single Sums		Uniform Series				Gradient Series	
	To Find F Given P (F|P,i%,n)	To Find P Given F (P|F,i%,n)	To Find F Given A (F|A,i%,n)	To Find A Given F (A|F,i%,n)	To Find P Given A (P|A,i%,n)	To Find A Given P (A|P,i%,n)	To Find P Given G (P|G,i%,n)	To Find A Given G (A|G,i%,n)
n								
1	1.3000	0.7692	1.0000	1.0000	0.7692	1.3000	0.0000	0.0000
2	1.6900	0.5917	2.3000	0.4348	1.3609	0.7348	0.5917	0.4348
3	2.1970	0.4552	3.9900	0.2506	1.8161	0.5506	1.5020	0.8271
4	2.8561	0.3501	6.1870	0.1616	2.1662	0.4616	2.5524	1.1783
5	3.7129	0.2693	9.0431	0.1106	2.4356	0.4106	3.6297	1.4903
6	4.8268	0.2072	12.7560	0.0784	2.6427	0.3784	4.6656	1.7654
7	6.2749	0.1594	17.5828	0.0569	2.8021	0.3569	5.6218	2.0063
8	8.1573	0.1226	23.8577	0.0419	2.9247	0.3419	6.4800	2.2156
9	10.6045	0.0943	32.0150	0.0312	3.0190	0.3312	7.2343	2.3963
10	13.7858	0.0725	42.6195	0.0235	3.0915	0.3235	7.8872	2.5512
11	17.9216	0.0558	56.4053	0.0177	3.1473	0.3177	8.4452	2.6833
12	23.2981	0.0429	74.3270	0.0135	3.1903	0.3135	8.9173	2.7952
13	30.2875	0.0330	97.6250	0.0102	3.2233	0.3102	9.3135	2.8895
14	39.3738	0.0254	127.9125	7.818E-03	3.2487	0.3078	9.6437	2.9685
15	51.1859	0.0195	167.2863	5.978E-03	3.2682	0.3060	9.9172	3.0344
16	66.5417	0.0150	218.4722	4.577E-03	3.2832	0.3046	10.1426	3.0892
17	86.5042	0.0116	285.0139	3.509E-03	3.2948	0.3035	10.3276	3.1345
18	112.4554	8.892E-03	371.5180	2.692E-03	3.3037	0.3027	10.4788	3.1718
19	146.1920	6.840E-03	483.9734	2.066E-03	3.3105	0.3021	10.6019	3.2025
20	190.0496	5.262E-03	630.1655	1.587E-03	3.3158	0.3016	10.7019	3.2275
21	247.0645	4.048E-03	820.2151	1.219E-03	3.3198	0.3012	10.7828	3.2480
22	321.1839	3.113E-03	1.067E+03	9.370E-04	3.3230	0.3009	10.8482	3.2646
23	417.5391	2.395E-03	1.388E+03	7.202E-04	3.3254	0.3007	10.9009	3.2781
24	542.8008	1.842E-03	1.806E+03	5.537E-04	3.3272	0.3006	10.9433	3.2890
25	705.6410	1.417E-03	2.349E+03	4.257E-04	3.3286	0.3004	10.9773	3.2979
26	917.3333	1.090E-03	3.054E+03	3.274E-04	3.3297	0.3003	11.0045	3.3050
27	1.193E+03	8.386E-04	3.972E+03	2.518E-04	3.3305	0.3003	11.0263	3.3107
28	1.550E+03	6.450E-04	5.164E+03	1.936E-04	3.3312	0.3002	11.0437	3.3153
29	2.015E+03	4.962E-04	6.715E+03	1.489E-04	3.3317	0.3001	11.0576	3.3189
30	2.620E+03	3.817E-04	8.730E+03	1.145E-04	3.3321	0.3001	11.0687	3.3219
36	1.265E+04	7.908E-05	4.215E+04	2.372E-05	3.3331	0.3000	11.1007	3.3305
42	6.104E+04	1.638E-05	2.035E+05	4.915E-06	3.3333	0.3000	11.1086	3.3326
48	2.946E+05	3.394E-06	9.821E+05	1.018E-06	3.3333	0.3000	11.1105	3.3332
54	1.422E+06	7.032E-07	4.740E+06	2.110E-07	3.3333	0.3000	11.1110	3.3333
60	6.864E+06	1.457E-07	2.288E+07	4.370E-08	3.3333	0.3000	11.1111	3.3333
66	3.313E+07	3.018E-08	1.104E+08	9.054E-09	3.3333	0.3000	11.1111	3.3333
72	1.599E+08	6.253E-09	5.331E+08	1.876E-09	3.3333	0.3000	11.1111	3.3333
120	4.712E+13	2.122E-14	1.571E+14	6.367E-15	3.3333	0.3000	11.1111	3.3333
180	3.234E+20	3.092E-21	1.078E+21	9.275E-22	3.3333	0.3000	11.1111	3.3333
360	1.046E+41	9.559E-42	3.487E+41	2.868E-42	3.3333	0.3000	11.1111	3.3333

Appendix B
M&V Guidelines

IPMVP

The International Performance Measurement and Verification Protocol (IPMVP) is the widely referenced framework for 'measuring' energy or water savings. It is especially used in energy performance contracts where savings must be reported to a client and may form the basis of a payment to an ESCO. IPMVP presents common terminology and defines full disclosure, to support rational discussion of often contentious M&V issues. It documents the state of the art, but does not specify project design since it is a high level framework. An M&V engineer is still needed to apply IPMVP principles to the 'measurement' of savings for each energy efficiency project.

A primary purpose of IPMVP is to publish current good M&V practice, as reassurance for the public about savings reports. For example, it explains the need to adjust raw differences in energy use for changes in conditions between baseline and savings reporting periods. Performance contracting industry growth in the USA was facilitated by the publication of the IPMVP. Its global use has similarly helped the EPC industry worldwide. IPMVP is freely available at www.evo-world.org under the Products tab. The latest edition of IPMVP is available in French and English, with older editions in 10 languages. IPMVP has been cited in many places, as listed on EVO's website under the Resources tab. EVO trains and certifies people with appropriate qualifications and proven knowledge of IPMVP, as described under the Services tab. EVO also provides other resources for the M&V community.

IPMVP is prepared in three Volumes:

Volume I Concepts and Options for
Determining Energy and Water Savings
Volume I defines terminology and suggests good practices for documenting the effectiveness of energy or water efficiency projects that are implemented in buildings and industrial facilities.

305

These terms and practices help managers to prepare M&V Plans, which specify how savings will be measured for each project. The successful M&V Plan enables verification by requiring transparent reports of actual project performance.

Volume II Indoor Environmental Quality (IEQ) Issues

Volume II reviews IEQ issues as they may be influenced by an energy efficiency project. It highlights good project design and implementation practices for maintaining acceptable indoor conditions under an energy efficiency project. It advises on means of measuring IEQ parameters to substantiate whether indoor conditions have changed from the conditions of the baseline when determining savings.

Volume III Applications

Volume III contains specific application guidance manuals for Volume I. The two current applications manuals address new building construction (Part I) and page 2 IPMVP Overview renewable energy additions to existing facilities (Part II). This Volume is expected to be an area of continued development as more specific applications are defined, or country-specific sections are contributed.

FEMP

If you operate a federal facility, you may find this guideline useful, although it references the IPMVP frequently. Still another great resource: http://www1.eere.energy.gov/femp/pdfs/mv_guidelines.pdf.

ASHRAE

Although not widely used, ASHRAE also has a guideline for M&V, which was being updated at the time of this printing

Appendix C

Energy Auditing Basics

Energy audits can mean different things to different individuals. The scope of an energy audit, the complexity of calculations, and the level of economic evaluation are all issues that may be handled differently by each individual auditor and should be defined prior to beginning any audit activities. This chapter will review the various approaches to energy auditing and outline a standard approach to organizing and conducting an energy audit.

An energy audit can be simply defined as a process of determining the types and costs of energy use in the building, evaluating where a building or plant uses energy, and identifying opportunities to reduce consumption

There is a direct relationship to the cost of the audit, how much data will be collected and analyzed, and the number of conservation opportunities identified. Thus, a first distinction is made between cost of the audit which determines the type of audit to be performed. The second distinction is made between the type of facility. For example, a building audit may emphasize the building envelope, lighting, heating, and ventilation requirements. On the other hand, an audit of an industrial plant emphasizes the process requirements.

TYPES OF ENERGY AUDITS

Before starting the energy audit, it is helpful to have some idea of the scope of the project and level of effort necessary to meet expectations. There are four basic types or levels of energy audit, any of which may meet your requirements.

The basic audit levels, in order of increasing complexity are:

Type 0—The Benchmarking Audit
This audit includes performing a detailed preliminary analysis of

energy use and costs, and determining benchmark indices like Btu per square foot per year and dollars of energy cost per square foot per year, based on utility bills. Very cost effective for multiple facilities. The EPA/ DOE EnergyStar Portfolio Manager is one of the best and easiest tools to use—and it is free.

Type I—The Walk-through Audit

The walk-through audit, as its name implies, is a tour of the facility to visually inspect each of the energy using systems. It will typically include an evaluation of energy consumption data to analyze energy use quantities and patterns as well as provide comparisons to industry averages or benchmarks for similar facilities. It is the least costly audit but can yield a preliminary estimate of savings potential and provide a list of low-cost savings opportunities through improvements in operational and maintenance practices. The level one audit is also an opportunity to collect information for a more detailed audit later on if the preliminary savings potential appears to warrant an expanded scope of auditing activity.

Type II—Standard Audit

The standard audit goes on to quantify energy uses and losses through a more detailed review and analysis of equipment, systems, and operational characteristics. This analysis may also include some on-site measurement and testing to quantify energy use and efficiency of various systems. Standard energy engineering calculations are used to analyze efficiencies and calculate energy and costs savings based on improvements and changes to each system. The standard audit will also include an economic analysis of recommended conservation measures.

Type III—Computer Simulation

The level three audit will include more detail of energy use by function and a more comprehensive evaluation of energy use patterns. This is accomplished through use of computer simulation software. The auditor will develop a computer simulation of building systems that will account for weather and other variables and predict year-round energy use. The auditor's goal is to build a base for comparison that is consistent with the actual energy consumption of the facility. After this baseline is built, the auditor will then make changes to improve efficiency of various systems and measure the effects compared to the baseline. This method also accounts for interactions between systems to help prevent overestimation of savings. Because of the time involved in collecting detailed equipment information, operational data, and setting up an accurate computer model, this is the most expensive level of energy audit but may be warranted if the facility or systems are more complex in nature.

ASHRAE DEFINITIONS OF ENERGY AUDITS
(Ref. Procedures for Commercial Buildings Energy Audits, 2005 ASHRAE)

ASHRAE has formalized a set of energy audit definitions that are widely used by energy auditors. The source of the following definitions is the 2005 ASHRAE Handbook of HVAC Systems.

> "**Preliminary Energy Use Analysis.** This involves analysis of historic utility use and cost and development of the energy utilization index (EUI) of the building. Compare the building's EUI to similar buildings to determine if further engineering study and analysis are likely to produce significant energy savings."

We have also identified this as a Benchmark Audit.

> "**Level I: Walk-through Analysis.** This assesses a building's current energy cost and efficiency by analyzing energy bills and briefly surveying the building. The auditor should be accompanied by the building operator. Level I analysis identifies low-cost/no-cost measures and capital improvements that merit further consideration, along with an initial estimate of costs and savings. The level of detail depends on the experience of the auditor and the client's specifications. The Level I audit is most applicable when there is some doubt about the energy savings potential of a building, or when an owner wishes to establish which buildings in a portfolio have the greatest potential savings. The results can be used to develop a priority list for a Level II or III audit."

• Also known as the "one-day" or "walk-through" audit, this approach involves a cursory analysis of energy bills and a brief survey of the building to produce a rough estimate of how efficiently energy is used in the building

- This level of effort will detect at least some of the "low-hanging fruit" and may suggest other options worthy of more study, but should never be viewed as comprehensive.

- Although this option is easiest, it also produces the crudest results, so don't be tempted into thinking you're done once you do this much—you've really only gotten started.

   "Level II: Energy Survey and Analysis. This includes a more detailed building survey and energy analysis, including a breakdown of energy use in the building, a savings and cost analysis of all practical measures that meet the owner's constraints, and a discussion of any effect on operation and maintenance procedures. It also lists potential capital-intensive improvements that require more thorough data collection and analysis, along with an initial judgment of potential costs and savings. This level of analysis is adequate for most buildings."

- By investing more effort in the building survey and energy analysis, and by adding some system performance testing, this method provides a breakdown of how energy is used in the building as well as a broader range of savings options, including simple capital investments.

- It accounts for the "people factor" and its effect on uncertainty of savings, and also explores maintenance procedures and assesses any impacts savings measures may have on them.

- Many facilities will find this level of analysis to be sufficient.

   "Level III: Detailed Analysis of Capital-intensive Modifications. This focuses on potential capital-intensive projects identified during Level II and involves more detailed field data gathering and engineering analysis. It provides detailed project cost and savings information with a level of confidence high enough for major capital investment decisions.

   The levels of energy audits do not have sharp boundaries. They are general categories for identifying the type of information that can be expected and an indication of the level of confidence in the results. In a complete energy management program, Level II audits should be performed on all facilities.

A thorough systems approach produces the best results. This approach has been described as starting at the end rather than at the beginning. For example, consider a factory with steam boilers in constant operation. An expedient (and often cost-effective) approach is to measure the combustion efficiency of each boiler and to improve boiler efficiency. Beginning at the end requires finding all or most of the end uses of steam in the plant, which could reveal considerable waste by venting to the atmosphere, defective steam traps, uninsulated lines, and lines through unused heat exchangers. Eliminating end-use waste can produce greater savings than improving boiler efficiency.

A detailed process for conducting audits is outlined in ASHRAE (2004)."

• Even more detailed data are gathered from field equipment. Extensive test measurements are taken which may include spot measurements and short-term energy monitoring. Possible risks are assessed, and intensive engineering and economic analysis produces reliable estimates of project energy and financial performance with the high confidence needed for major capital projects.

• Although not defined by ASHRAE as an investment grade audit, it is often called this by many energy auditors.

• This analysis digs into the details of any large capital projects you may be considering as a result of previous, simpler audits. AEE, as well as many others, requires computer simulation to be part of an investment grade audit.

• These audit approaches tend to overlap in practice. All three assess the potential energy savings and initial cost of various energy savings strategies, so in that sense all are similar. The differences are in your confidence that you've truly found all your savings opportunities, the accuracy of the expected savings and initial cost, and how much information you have about the difficulty of the project implementation and the likely persistence of the savings over time. The devil is definitely in the details.

- All level II and level III audits involve collecting general building data (location, size, usage type, energy sources), historical energy use data, and energy systems data (type of equipment in the envelope, lighting, HVAC, service water, etc.) to get a description of the facility. The more detailed the available data are, the more complete this description can be. For example, submetering within a building makes it easy to call out specific end uses or facility areas, and having daily or even hourly consumption data allows you to call out time patterns normally buried within the monthly billing cycle.

- All these data then feed an energy use analysis that lays out how much energy is consumed for each major end use in the building, such as space heating, space cooling, lighting, air distribution, etc. This defines a baseline scenario for future years, is no energy projects are undertaken. A similar analysis can be done with respect to peak energy demand

- If you're serious about saving as much energy cost as possible with the quickest payback time and least hassle, take the time to plan your energy projects right. Perform a good energy audit, and assess its results carefully based on the needs of your facility, whether based on annual savings, initial cost, payback time, synergistic comfort benefits to occupants, or recurring maintenance hassle. The rewards are well worth the work.

THE INVESTMENT GRADE ENERGY AUDIT

In most facilities, companies, and other corporate settings, upgrades to a facility's energy infrastructure must compete for capital funding with non-energy-related investments. Both energy and non-energy investments are commonly rated on a standard set of financial criteria that generally stress the expected return on investment (ROI) and often the life cycle costs. The projected operating savings from the implementation of energy projects must be developed such that they provide a high level of confidence. In fact, investors often demand guaranteed savings. The investment-grade audit expands on the detailed audit Levels II and III above, and relies on complete engineering studies in order to detail technical and economic issues necessary to justify the investment related to the transformations. In most cases, detailed hour-by-hour computer simulation modeling will be required.

The formal AEE description of the investment grade energy audit is:

Investment Grade Audit

- This audit includes weighing financial risk into the economic calculations of a type II or III energy audit.
- It will often include computer simulation and enhanced financial analysis tools such as life cycle costing. Additional requirements may be specified by individual clients.
- This audit can be utilized to obtain funding for the projects identified.

The Association of Energy Engineers has developed three certification programs for professionals practicing energy auditing: The Certified Energy Auditor (CEA) Program, The Master's Level CEA Certification Program, and the Certified Energy Manager (CEM) Program.

THE CERTIFIED ENERGY AUDITOR (CEA) PROGRAM FOR PROFESSIONAL CERTIFICATION

The Mark of an Energy Professional

In 2006, the Certified Energy Auditor (CEA) and Certified Energy Auditor in Training (CEAIT) certifications were developed and added to the impressive portfolio of certifications offered by the Association of Energy Engineers. Rising energy costs and inefficiency in plants and buildings is continually driving the need for trained and experienced energy auditors. The CEA certification is one that identifies professionals as having the required knowledge and experience needed to succeed in the growing field of energy auditing.

Objectives

- To raise the professional standards of those engaged in energy auditing.

- To improve the practice of energy auditors by encouraging energy auditing in a continuing education program of professional development.

- To identify persons with acceptable knowledge of the principles and practices of energy auditing through completing an examination and fulfilling prescribed standards of performance and conduct.

- To award special recognition to those energy auditing professionals who have demonstrated a high level of competence and ethical fitness in energy auditing.

The Certified Energy Manager (CEM) Program for Professional Certification

CEM®

When you've earned the right to put the initials "CEM" behind your name, you've distinguished yourself among energy management professionals. Simply put, the designation CEM, which stands for Certified Energy Manager, recognizes individuals who have demonstrated high levels of experience, competence, ,proficiency, and ethical fitness in the energy management profession. By attaining the status of CEM, you will be joining an elite group of 6,000 professionals serving industry, business and government throughout the U.S. and in 22 countries abroad. These high-achieving individuals comprise a "Who's Who" in the energy management field.

The Master's Level Certified Energy Auditor (CEAM) Program

New for 2012 and developed with grant funding from the US Department of Energy, the Master's Level Certified Energy Auditor (MCEA) certification is designed to reach beyond the typical equipment replacements and develop a plan which considers additional areas of energy such as indoor air quality, code compliance, operation and maintenance, risk mitigation, commissioning, and investment grade details.

Objectives
- To raise the professional standards of those engaged in energy auditing.

- To improve the practice of energy auditors by encouraging energy auditing in a continuing education program of professional development.

- To identify persons with acceptable knowledge of the principles and practices of energy auditing through completing an examination and fulfilling prescribed standards of performance and conduct.

- To award special recognition to those energy auditing professionals who have demonstrated a high level of competence and ethical fitness in energy auditing.

The Mark of An Energy Professional

Since its inception in 1981, the Certified Energy Manager (CEM®) credential has become widely accepted and used as a measure of professional accomplishment within the energy management field. It has gained industry-wide use as the standard for qualifying energy professionals both in the United States and abroad. It is recognized by the U.S. Department of Energy, the Office of Federal Energy Management Programs (FEMP), and the U.S. Agency for International Development, as well as by numerous state energy offices, major utilities, corporations and energy service companies.

THE AUDIT PROCESS

The first step in any energy audit should be to collect energy bills and perform a benchmark audit. Once you have established the level of actual audit to be performed, you can begin collecting information on the structural and mechanical components that affect building energy use, and about the operational characteristics of the facility. Much of this information can and should be collected prior to the actual site-visit. A thorough evaluation of energy use and systems before going on-site will help identify areas of savings potential and help make best use of your on-site time.

An organized approach to auditing will help you collect useful information and reduce the amount of time spent evaluating your facility. By splitting the audit process into three distinct components, *pre-site work, the site visit, and post-site work,* it becomes easier to allocate your time for each step and leads to a more comprehensive and useful audit report.

The following sections describe the tasks associated with each step of the audit process.

PRE-SITE WORK

Pre-site work is important in getting to know basic aspects of the building. This preparation will help ensure the most effective use of your on-site time and minimize disruptions to building personnel.

A thorough pre-site review will also reduce the time required to complete the on-site portion of the audit. The pre-site review of building systems and operation should generate a list of specific questions and issues to be discussed during the actual visit to the facility.

Pre-site Tasks

1) Collect and review one to two years of utility energy data. Tabulate and graph the data. Check for seasonal patterns, unusual spikes, and accuracy of the billings. Graphing consumption and cost data makes it easier to understand how each building uses energy. By determining seasonal and base loads, then apportioning energy use among specific building systems such as heating, cooling, lighting and hot water, it becomes easier to identify areas with the greatest savings potential. It's also important to include electric demand kilowatts and demand charges in your evaluation. (See Energy Accounting section for additional information on utility bill analysis.) Pie charts of energy use and cost by fuel type can offer compelling documentation of overall energy uses and expenses.

2) Obtain mechanical, architectural, and electrical drawings and specifications for the original building as well as for any additions or remodeling work that may have been done. Try the local building department or original architect if the owner doesn't have them. If any energy audits or studies have been done in the past, obtain a copy and review them.

3) Draw a simple floor plan of the building on 8-1/2 × 11 or 11 × 17 inch paper. Make several copies to use for taking notes during the actual site visit. Use separate copies for noting information on locations of HVAC equipment and controls, heating zones, light levels and other energy related systems.

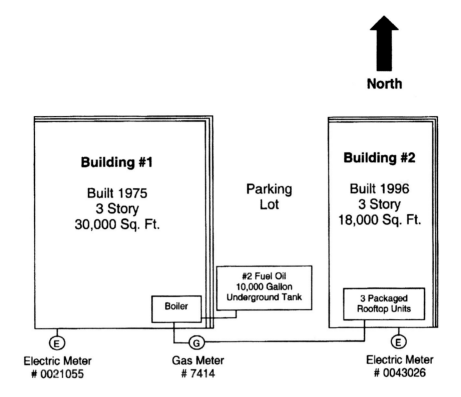

4) Calculate the gross square footage of conditioned space using outside building dimensions multiplied by the number of stories. Substantial areas that are not conditioned and occupied can be subtracted from the gross square footage.

5) Use audit data forms to collect, organize and document all pertinent building and equipment data. Audit workbooks containing checklists, equipment schedules, and other forms are available from a variety of sources including US Department of Energy, EnergyStar, ASHRAE,and your State Energy Office. You may also find it useful to develop your own forms to meet your specific needs. To save time, fill out as much of the form as possible using the building plans and specifications before starting on-site work.

6) Develop a building profile narrative that includes age, occupancy, description, and existing conditions of architectural, mechanical,

and electrical systems. Note the major energy consuming equipment or systems.

7)   Calculate the energy use index (EUI) in Btu/sq ft/year and compare it with EUIs of similar building types using the chart in the energy accounting section. The EUI is calculated by converting annual consumption of all fuels to Btus then dividing by the gross square footage of the building. It can be a good indicator of the relative potential for energy savings. A comparatively low EUI indicates less potential for large energy savings.

A great, free resource is available from EnergyStar. The Portfolio Manager is an easy-to-use, free benchmarking program available from EnergyStar.gov. This process is called a benchmark audit.

While completing your pre-site review, note areas of particular interest and write down any questions you may have in advance. Typical questions may concern lighting type and controls, HVAC zone controls or morning warm-up operation. Other questions may be on maintenance practices for pieces of equipment you've identified which typically require regular servicing.

If you are auditing a building other than one you operate, obtain the data discussed above and confirm your preliminary observations with the building manager or operator by phone prior to your visit. Ask them if they are interested in particular conservation projects or planning changes to the building or its systems. Try to schedule the audit at a time when the systems you want to check are in operation and arrange to have the building operator accompany you during the site visit.

Develop a list of potential energy conservation measures (ECMs) and operation and maintenance (O&M) procedures as you conduct this preliminary research. Your state energy office or local utility companies should be able to provide you with more information on conservation technologies and O&M recommendations. If you do some homework first, you will be better able to discuss energy saving measures with the building manager.

Develop a Site Sketch
Prepare a site sketch of the building or complex which shows the following information:
•   Relative location and outline of the building(s).

- Name and building number of each building. (Assign building numbers if none exist.)
- Year of construction of each building and additions.
- Square footage of each building and additions.
- Location, fuel type and I.D. numbers of utility meters.
- Areas served by each utility meter.
- Location of heating and cooling plants and equipment.
- North orientation arrow.

The Site Visit

With pre-site work completed, you should have a basic understanding of the building and its systems. The site visit will be spent inspecting actual systems and answering specific questions from your pre-site review.

Plan to spend at least a full day on-site for each building. The amount of time required will vary depending on the completeness of the pre-site information collected, the complexity of the building and systems, and the need for testing of equipment. Small buildings may take less time. Larger buildings can take two days or more.

Here are some steps to help you conduct an effective audit:

- Have all necessary tools available on site. Try to anticipate basic hand tools and test equipment you will need to perform a thorough inspection. Some basic audit tools you'll want to bring along include:

  —Notebook

  —Calculator

  —Flashlight

  —Tape Measure

  —Pocket Thermometer

  —Light Meter

  —Pocket Knife

  —Camera

  —Binoculars

  —Mini Data Loggers

A more detailed description of specialized audit instrumentation is included in Chapter 4.

- Prior to touring the facility, sit down with the building manager to review energy consumption profiles and discuss aspects of the facility you aren't able to see such as occupancy schedules, operation and maintenance practices, and future plans that may have an impact on energy consumption.

- Confirm the floor plan on your drawing to the actual building and note major changes. Use copies of the floor plan to note equipment locations such as boilers, chillers, DHW heaters, kitchen appliances, exhaust fans, etc., as well as lighting types, levels, and switching, photo locations, room temperatures, general conditions and other observations.

- Fill out the audit data sheets. Use them to organize your site visit and as a reminder to collect information missing from pre-site documents.

- Look at the systems relating to the ECMs and O&Ms on your preliminary list. Review the application of your recommendations and note any problems that may affect implementation. Add additional measures to your list as you tour the facility.

- Take pictures as you walk through the building. Include mechanical equipment, lighting, interior workspaces, common areas and halls, and the exterior including the roof. They are useful in documenting existing conditions, discussing problems and issues with colleagues, as well as serving as a reminder of what you inspected. Building managers will find them useful for explaining conservation measures to administrators and building occupants.

- Take basic measurements of light levels, temperature, relative humidity, and voltages.

POST-SITE WORK

Post-site work is a necessary and important step to ensure the audit will be a useful planning tool. The auditor needs to evaluate the information gathered during the site visit, research possible conservation opportunities, organize the audit into a comprehensive report, and make recommendations on mechanical, structural, operational and maintenance improvements.

Post-site work includes the following steps:

- Immediately after the audit, review and clarify your notes. Complete information you didn't have time to write down during the audit. Use copies of the floor plan to clean up notes for permanent records.

- Review and revise your proposed ECM and O&M lists. Eliminate those measures lacking potential and document why they were eliminated. Conduct preliminary research on potential conservation measures and note conditions that require further evaluation by an engineer or other specialist.

- Process your photos and paste or import pictures on 8-1/2 × 11 inch pages. Number the photographs and note on a floor plan the location where each photo was taken. Identify and add notes under the pictures as needed.

- Organize all charts, graphs, building descriptions, audit data sheets, basic measurements, notes and photos into a 3 ring binder. Energy auditing can be an ongoing process. By keeping all building information in a dedicated binder or file, records can be easily added or updated and can be very useful to architects and engineers if future work is done on the building.

THE AUDIT REPORT

The general flow of audit activities is to identify all energy uses and costs, identify energy systems, evaluate the condition of the systems, analyze the impact of improvements to those systems, and write up an energy audit report. This report explains the existing conditions of the building(s) in terms of the envelope, equipment, lighting, and occupancy, followed by recommendations to improve efficiency through improvements in operation and maintenance items, or (O&Ms), and through installation of energy conservation measures, or ECMs.

Effectively communicating audit findings and recommendations increases the chance of action being taken to reduce energy consumption. When preparing the audit report, keep in mind the various audiences that will be using each section and try to customize each section to most effectively reach that audience.

Typical audiences for audit reports include:
- CEO, COO, Administrator, Superintendent
- Facilities and Plant Managers
- CFO, Controller
- Plant Engineer
- Operations and Maintenance Staff

The following outlines the basic components of a well-organized audit report:

I.   Executive Summary

The Executive Summary should be a simple, straight forward and to the point explanation of the current situation, recommended improvements, and advantages of taking recommended actions. Include a brief introduction to the facility and describe the purpose of the audit and overall conclusions. An executive may read no further than this one or two-page introduction so make sure that you have expressed very clearly what specific actions you want them to take.

II.   Building Information

This section provides a general background of the facility, its function, its mechanical systems, and operational profile. It should include a description of the building envelope, age and construction history, operating schedules, number of employees and occupancy patterns, and a discussion of the operation and maintenance program. It is also useful to include a floor plan, selected photos of the facility and mechanical systems, a description of energy types used in the plant, and a description of the primary mechanical systems and controls.

III.   Utility Summary

Energy accounting information for the last one to two years is included in this section. Attach selected charts and graphs that were developed for analysis that are easy to understand and demonstrate the overall consumption patterns of the facility. Choose the information for each graph to suit each target audience. For example, actual monthly consumption by electricity and fuel type may be of more interest to the engineering and maintenance staff while annual costs or dollar-savings information may be more appropriate for administrative personnel. Pie charts of energy use and cost by fuel type can offer compelling documentation of overall energy uses and expenses. For electricity, include data on both energy use and peak demand.

Of increasing importance is examining the use of water and sewer in facilities. Reducing the use of water and sewer is a requirement for many facilities including federal facilities.

Include a summary of overall facility benchmarks, energy use indices, and comparisons with industry averages.

You may also want to include a copy of the utility rate schedules and any discussion or evaluation of rate alternatives for which the facility may qualify.

IV.   Energy Conservation Measures (ECMs)

Begin this section with a summary list of Energy Conservation Measures that meet the financial criteria established by the facility owner or manager. For each measure, include the measure name, estimated cost, estimated savings, and simple payback in a summary chart. A one or two page description of each energy conservation measure and support calculations should follow this summary chart. Include the ECM description, energy use and savings calculations, and the simple payback, rate of return, and net present value or life cycle cost analysis. It's also a good idea to discuss any assumptions that were made regarding operation or equipment efficiency. ECMs that were considered but fell out of current financial criteria should also be listed and identified as have been evaluated.

V.   Operation and Maintenance Measures (O&Ms)

This section will address operational and maintenance issues observed during the site visit. Include descriptions of specific low-cost operational and maintenance items that require attention. Include items that will reduce energy consumption and costs, address existing problems, or improve practices that will help prolong equipment life of systems not being retrofit. It is also useful to the owner to include cost and savings estimates of O&M recommendations.

VI. Appendices

Support material and technical information not included elsewhere in the report can be added to the appendices. Typical information in this section includes, floor plans and site notes as appropriate, photos, audit data forms, motor, equipment, and lighting inventories, and equipment cut sheets of existing or recommended systems.

SUMMARY

When you've completed your audit activities, you should have a good understanding of the primary drivers affecting facility energy use. By identifying the energy consuming components of a building or plant and documenting the existing conditions, conservation opportunities can be identified and prioritized. Set up a meeting with the building or plant manager to go over your report. Discuss your recommendations for conservation actions, methods of funding ECMs. Include training recommendations for building operators and occupants that will improve the operating efficiency of the building as well as training required for maintenance and operation of newly install measures.

Appendix D

Project Development Agreement

CUSTOMER'S PREMISES: XYZ PLANT

Executive Summary:

ESCO proposes to conduct a feasibility study (aka project development agreement) to determine the costs and savings for a project bundle that includes the following productivity improvement measures:

1. Glass Thickness Monitoring and Statistical Control
2. Variable Speed Drives on Fans
3. Install Water Well for Cooling Tower Make-Up
4. Compressed Air Improvements

In addition, ESCO will conduct a holistic preliminary assessment of different options to improve furnace operations. These include Oxifuel Firing, Temperature Modulation, Combustion Efficiency Modulation and Waste Heat Recovery.

CLIENT (Customer) will not need to pay for the Feasibility Study if ESCO cannot develop a project bundle that has a simple payback less than two years (after-taxes) OR can be financed for three years such that the annual savings are greater than the associated finance payments (aka "budget neutral"). Within 75 days (from signing this document), ESCO will develop the project bundle and present it to CLIENT. If ESCO meets the criteria mentioned above, ESCO expects to implement the projects for CLIENT. The cost for the feasibility study will be the lesser of $300,000 or 10% of the bundled project price. When CLIENT implements the turnkey projects with ESCO, the actual development costs will be included as part of the project costs.

EVALUATION STUDY

ESCO agrees to undertake a detailed evaluation study of the Customer's Premises identified above to determine the energy consumption and operational characteristics of the Premises and to identify the energy conservation measures, procedures, and other services that could be provided by ESCO in order to reduce the Customer's energy consumption and operating costs on the Premises. Customer agrees to provide its complete cooperation in the conduct and completion of the study. ESCO will provide to the Customer a written report within 75 days of the effective date of this Agreement. The report will include:

(a)   a list of specific energy conservation measures that ESCO proposes to install;

(b)   a description of the operating and maintenance procedures that ESCO believes can reduce energy consumption and operating costs at the Premises; and

(c)   an estimate of the energy and operating costs that will be saved by the equipment and procedures recommended in the report.

ENERGY USAGE RECORDS AND DATA

During the evaluation study, Customer will furnish to ESCO, upon its request, accurate and complete data concerning energy usage and operational expenditures for the Premises, including the following data for the most recent three years from the effective date of this Agreement:

• actual utility bills supplied by the utility and other relevant utility records;

• occupancy and usage information;

• descriptions of any changes in the building structure or its heating, cooling, lighting, or other systems or their energy requirements;

• descriptions of all energy-consuming or energy-saving equipment used on the Premises;

- descriptions of energy management and other relevant operational or maintenance procedures utilized on the Premises;

- summary of expenditures for outsourced maintenance, repairs, or replacements on the Premises;

- copies of representative current tenant leases, if any; and

- prior energy audits or studies of the Premises, if any.

PREPARATION OF PERFORMANCE CONTRACTING PROJECT AGREEMENT

Within 30 days after the submission to Customer of the report described under paragraph 1 of this Agreement, ESCO will prepare and submit to the Customer a Performance Contracting Project Agreement to implement the energy conservation measures, procedures, and services identified in the report that could reduce the Customer's energy consumption in the Premises. This Project Agreement shall include ESCO's Performance Contract and appropriate Schedules, copies of which are provided in connection with this Schedule 4, Project Development Agreement Schedule.

PRICE AND PAYMENT TERMS

Customer agrees to pay to ESCO the sum of $ 300,000 or 10% of final project bundle price (whichever is less) within 30 days after the delivery to the Customer of the report described under paragraph 1 of this Agreement. However, Customer will have no obligation to pay this amount if:

(a) ESCO cannot develop a project bundle that has a simple payback of less than two years (after-taxes) OR can be financed for three years such that the annual savings are greater than the associated finance payments (aka "budget neutral").

INDEMNITY

ESCO and the Customer agree that ESCO shall be responsible only for such injury, loss, or damage caused by the intentional misconduct or the negligent act or omission of ESCO. ESCO and the Customer agree to indemnify and to hold each other, including their officers, agents, directors, and employees, harmless from all claims, demands, or suits of any kind, including all legal costs and attorney's fees, resulting from the intentional misconduct of their employees or any negligent act or omission by their employees or agents. Neither ESCO nor the Customer will be responsible to the other for any special, indirect, or consequential damages.

DISPUTES

If a dispute arises under this Agreement, the parties shall promptly attempt in good faith to resolve the dispute by negotiation. All disputes not resolved by negotiation shall be resolved in accordance with the Commercial Rules of the American Arbitration Association in effect at that time, except as modified herein. All disputes shall be decided by a single arbitrator. A decision shall be rendered by the arbitrator no later than nine months after the demand for arbitration is filed, and the arbitrator shall state in writing the factual and legal basis for the award. No discovery shall be permitted. The arbitrator shall issue a scheduling order that shall not be modified except by the mutual agreement of the parties. Judgment may be entered upon the award in the highest state or federal court having jurisdiction over the matter. The prevailing party shall recover all costs, including attorney's fees, incurred as a result of the dispute. If the Customer is a state or local governmental entity, then this paragraph shall not apply.

MISCELLANEOUS PROVISIONS

This Agreement cannot be assigned by either party without the prior written consent of the other party. This Agreement is the entire Agreement between ESCO and the Customer with respect to development activities and services and supersedes any prior oral understandings, written agreements, proposals, or other communications between ESCO and the Customer. Any change or modification to this Agreement will not be effective unless made in writing. This written instrument must specifically indicate that it is an amendment, change, or modification to this Agreement.

To the extent that ESCO and the Customer have entered into a Performance Contracting Project Agreement, this Schedule is attached to and made a part of the Performance Contract between ESCO and the Customer, dated _____.

CLIENT (CUSTOMER):   **ESCO, INC.**

Signature: ——————————   Signature:——————————

Printed Name: ——————————   Printed Name: ——————————

Title: ——————————   Title: ——————————

Appendix E

AGREEMENT FOR ENERGY CONSERVATION AND DEMAND SIDE MANAGEMENT SERVICES BETWEEN

THE UNITED STATES OF AMERICA AND

_____ UTILITY COMPANY

This Agreement for implementation of Energy Conservation Measures (ECMs) is entered into this _____ day of _____, 200_, by and between _____ Utility Company (Utility) and the United States of America (Government), represented by the Contracting Officer executing this Agreement. The signatories to this Agreement will be sometimes collectively referred to as the "Parties" and individually as a "Party." This Agreement (when signed by the Parties), any Task Orders (T.O.) executed pursuant to this Agreement, and any other associated agreements shall constitute the entire Contract between the Parties with respect to a particular ECM. A term or condition contained in this Agreement may be amended at any time by mutual written agreement of the Parties. However, termination, modification, or expiration of a term or condition shall not retroactively affect T.O.s previously entered into under this Agreement.

The Parties agree to the following principals, concepts and procedures:

GENERAL CONDITIONS

GC.1 Purpose. The Government desires assistance in accomplishing ECMs at _____ Installation ("Installation") (may substitute "at all Installations within the Utility Company's service area, to include [list the installations by name] ("hereinafter, "Installations")). The purpose of this Agreement is to facilitate the implementation of ECMs through T.O.s. This Agreement sets forth the terms and conditions under which subsequent T.O.s may be entered into between the Parties.

GC.2 Definitions. Terms used in this Agreement shall have the following definitions:

Acceptance— Written acceptance by the authorized representative of the Government of an individual Phase or completed ECM pursuant to a T.O.

Carrying Charge— For the purpose of this Agreement, Carrying Charge shall be an interest rate applied to all ECM Costs incurred by the Utility until permanent financing is put in place or the Government pays the ECM Cost. Accrued interest shall be considered an ECM Cost.

Contracting Officer— A Government official authorized to enter into, administer, and/or terminate a contract on behalf of the Government, and who is authorized to make related determinations and findings within the limits established pursuant to Government regulations.

Contracting Officer's Representative (COR) or Contracting Officer's Technical Representative (COTR)— A local or project site representative of the Contracting Officer delegated specific limited authority, as set forth in a formal delegation letter signed by the Contracting Officer, for a given T.O.

Energy Conservation Measure (ECM)— One or more ECPs completed, or to be completed, under a T.O. including the feasibility study, engineering and design, operation and maintenance, and/or implementation of one or more ECPs, which include, but are not limited to, energy and water conservation, energy efficient maintenance, energy management services, facilities alterations, and installation and maintenance of energy saving devices and technologies. ECMs should be readily available and demonstrate an economic return on investment, as required by Title 10 U.S.C., Section 2911.

Energy Conservation Measure Cost (ECM Cost)— The total cost may include, but is not limited to the Work, finance charges and overhead and profit, for the feasibility study, engineering and design, implementation and operation and maintenance of an ECM, less any financial incentive or rebates, if provided by the Utility. Payment for completed ECMs shall be calculated based upon the ECM Cost.

Energy Conservation Project (ECP)— A specific project intended and designed to provide any of the following: energy savings, demand reduction, efficiency improvements and water conservation. ECPs are described in more detail in Section GC 17.

Occupied Period—Hours during which a facility or building is occupied or used in the normal course of business.

Quality Assurance Evaluator (QAE)—A functionally qualified person who evaluates or inspects the **contractor's** performance of service in accordance with the quality assurance surveillance plan written specifically for the contracted service to be evaluated. The QAE performs technical monitoring of contractor actions, is responsible for requesting products and services through a government contract, and manages the day-to-day tasks of the contract.

Quality Control—A management function whereby control of quality of raw or produced material is exercised for the purpose of preventing production of defective material. For purposes of this Agreement, quality control is those actions taken by a *contractor* to control the production of outputs to ensure that they conform to the **contract** requirements.

Possession—When the Government takes *beneficial* occupancy of an ECP ("Possession of an ECP") or an ECM ("Possession of an ECM").
Subcontractor— Any corporation, partnership or individual hired directly by the Utility to perform a service or provide a product under this Agreement and T.O.s resulting from this Agreement.

Task Order (T.O.)—A binding contractual action entered into under this Agreement for the feasibility study, engineering and design, implementation, and/or operation and maintenance of, or any activity related to an ECM. (A T.O. can also be identified as a Delivery Order (D.O.).)

Termination Schedule—A schedule developed for each financed ECM specifying the lump sum payment necessary, at any time during the contract period following the initial Government payment, for the complete repayment of the ECM Costs, including any finance costs accrued to that point.

Work—All labor, materials, tools, equipment, services, transportation and/or other items required for the completion of the ECM.

GC.3 Term. This Agreement shall have a term of _____ years. The term may not exceed ten (10) years. This Agreement may be terminated in its entirety by either Party upon thirty (30) days written notice to the other Party. Thereafter, no new T.O.s shall be entered into under this Agreement. Termination, modification or expiration of this Agreement shall not affect in any way T.O.s previously entered into under this Agreement. This Agreement shall be effective from the date it is signed by both Parties. In the event the Parties sign this Agreement on different dates, then the effective date shall be the latter of the two dates.

GC.4 Services to be Provided by the Utility. The Utility shall provide preliminary audits, feasibility studies, engineering and design studies, and all initial capital, labor, material, supplies and equipment to identify, implement, operate or maintain ECMs in accordance with T.O.s entered into pursuant to this Agreement. These services may be ordered individually, as a group or in any combination under a single T.O.

GC.5 Information. Subject to national security constraints and unless otherwise prohibited by law, the Government shall provide the Utility with any information requested by the Utility to comply with regulatory commission requirements.

GC.6 Relationship of Parties. The Government acknowledges that the Utility and/or its Subcontractors shall each perform their work as independent contractors and the Government shall have no direct control and supervision of Utility or Subcontractor employees, who shall not be considered employees or agents of the Government for any purpose. The Utility, in negotiations with its Subcontractors, will ensure that the Government will be the direct beneficiary of any and all product and service guarantees and warranties.

7
. The Utility may perform some or all of the Work under a Task Order itself or through Subcontractors. When practical, the Utility shall competitively select Subcontractors for the purpose of determining the reasonableness of Subcontractor prices. When competition is not practical, price reasonableness may be determined by comparing proposed prices

with those obtained for the same or similar work, prices published in independent cost guides, published in competitive price lists or developed by independent sources.

Subcontractor selection shall be based on cost, experience, past perfrmance, reliability, and such other factors as the Utility may deem appropriate, as long as such factors are practicably related to the Government's *minimum needs*. In no event may such services be provided by Subcontractors listed as excluded from Federal Procurement Programs, which list is maintained by GSA pursuant to 48 C.F.R. 9.404. For any T.O., the Utility may submit the names of proposed Subcontractors to the Government Contracting Officer to ensure they are not excluded pursuant to 48 C.F.R. 9.404.

GC.8 Authority of Contracting Officer. The Government's Contracting Officer shall be the only Government official authorized to enter into and/or modify a T.O. entered into under this Agreement.

GC.9 Ownership of Work Product. The Government may elect not to use the Utility to implement the ECM. If the Government so elects, it will pay for any accepted work, including any equipment, completed studies, and engineering and design work. Title to any work done by the Utility for the Government under a T.O. shall become the property of the Government at the time of Acceptance of the Work.

GC.10 Responsibility for Operation and Maintenance. The operation and maintenance of the equipment installed pursuant to any T.O. executed under this Agreement shall be the responsibility of the Utility during the payment term unless otherwise provided in the T.O.

GC.11 Government Projects. The Government shall not be restricted from implementing equipment installation, construction projects and ECMs independent of work performed under this Agreement, including installing new energy conservation equipment, removing existing energy consuming equipment, or adding new energy consuming equipment. The Government will notify the Utility prior to implementing projects that may affect ECMs under this Agreement.

GC.12 ECM Performance Verification. Each T.O. shall include procedures that are mutually agreeable to the parties to verify ECM performance following installation.

GC.13 Emission Credits. All on site Government emission credits earned by virtue of T.O.s entered into hereunder shall be the property of the Government.

GC.14 Order of Precedence. The Government and Utility shall determine in this Agreement or subsequent T.O.s the precedence given to the T.O., this Agreement or other documents, exhibits and attachments in the event an inconsistency arises among these documents.

GC.15 Preliminary Audits. At the request of the Government or the Utility and upon the mutual consent of both parties, the Utility will conduct, at no cost to the Government, an audit consisting of an on-site building investigation and evaluation for a mutually agreeable facility to determine if any significant energy conservation opportunities exist and whether further detailed energy analysis is warranted. Government buildings/facilities plans will be made available upon request. Requests for plans shall be made to the COR at least fifteen (15) calendar days in advance of the audit start date. The Utility will provide a written report of the audit to the Government, **normally at no cost**. The Utility will utilize historical building data, utility data, and information obtained by the Utility to identify ECPs. Using this information, the Utility will generate a prioritized list of recommendations, in sequence of implementation, that are life-cycle cost effective and can be implemented in the facility being audited. The preliminary audit, to the extent applicable, shall include but not be limited to the following information:

(a)   Preliminary estimated energy and water savings,
(b)   Preliminary estimated cost savings, including reduced maintenance costs,
(c)   Current utility rates,
(d)   Preliminary retrofit cost,
(e)   Utility financial incentive/rebate, if any,
(f)   Description of existing equipment,
(g)   Description of the proposed retrofit equipment,
(h)   Overview of the general environmental impact and potential hazardous wastes identified through existing facility records, if any.

GC.16 ECM Proposal. After reviewing the preliminary audit, the Government may request a proposal from the Utility, for the evaluation of an ECM. The Utility shall submit an ECM proposal setting forth a prioritized list of the recommended ECPs within the ECM, a preliminary estimate of the cost to implement each ECP, the total costs for implementing the ECM (including estimated feasibility study, engineering and design, and implementation costs), and estimated cost savings.

GC.17 Energy Conservation Projects. The Utility may propose ECMs which include one or more ECPs. ECPs that substitute one energy type for another (e.g., natural gas in lieu of electricity) will not be considered for implementation unless a net overall energy or cost reduction can be demonstrated, based on current market energy prices. Potential ECPs include, but are not limited to:

(a)   Interior and exterior lighting replacement,
(b)   Transformer replacement,
(c)   Lighting control improvements,
(d)   Motor replacement with high efficiency motor,
(e)   Construction of alternative generation or cogeneration facilities,
(f)   Boiler control improvements,
(g)   Packaged air conditioning unit replacement,
(h)   Cooling tower retrofit,
(i)   Economizer installation,
(j)   Energy management control system (EMCS) replacement/alteration,
(k)   Occupancy sensors,
(l)   LED exit sign installation,
(m)   Fans and pump replacement or impeller trimming,
(n)   Chiller retrofit,
(o)   Upgrade of natural gas-fired boilers with new controls (low NO_x burners),
(p)   Solar domestic hot water system,
(q)   Solar air preheating system,
(r)   Steam trap maintenance and replacement,
(s)   Insulation installation,
(t)   Variable speed drive utilization,
(u)   Weatherization,
(v)   Window replacement,
(w)   Window coverings and awnings,
(y)   Reflective solar window tinting,
(z)   Fuel cell installation,
(aa)   Photovoltaic system installation,

(bb)   Faucet replacement (infrared sensor),

(cc)   Replacement of air conditioning & heating unit with a heat pump,

(dd)   Addition of liquid refrigerant pump to a reciprocating air conditioning unit,

(ee)   High efficiency refrigerator replacement,

(ff)   High efficiency window air conditioner replacement,

(gg)   Water conservation device installation (e.g., flow restrictors, low flow flush valves, waterless urinals, horizontal axis washing machines),

(hh)   Installation, maintenance and operation of power quality and reliability measures including UPS systems, back-up generators, emergency generators, etc.,

(ii)   Fuel switching technology,

(jj)   Infrared heating system,

(kk)   Heat pipe dehumidification,

(ll)   Flash bake commercial cooking,

(mm)   Thermal energy storage system,

(nn)   Operation, maintenance, modification and/or extension of utility distribution and collection system,

(oo)   Training that will result in reduced energy costs,

(pp)   Power factor correction measures and equipment,

(qq)   Installation, maintenance and operation of standby propane facility,

(rr)   Installation, maintenance and operation of gas distribution system and associated equipment,

(ss)   Water distribution system leak detection, and cost effective repair,

(tt)   Any other ECP that is cost effective using the then current DoD prescribed procedures and standards, and which encourages the use of renewable energy, reduces the Government's energy consumption or energy demand or results in other energy infrastructure improvements.

GC.17.1 ECM Restrictions. The Government shall not consider ECMs which include:

(a)   Measures which could jeopardize existing agency missions,

(b)   Measures which could jeopardize the operation of, or environmental conditions of, computers or computer rooms,

(c)   Unless waived by the Contracting Officer, measures that would result in increased water consumption (e.g., once-through fresh water cooling systems),

(d) Measures which would violate any federal, state, or local laws or regulations,

(e) Measures which degrade performance or reliability of existing Government equipment,

(f) Unless waived by the Contracting Officer, measures that would reduce energy capacity currently reserved for future growth, mobilization needs, safety, emergency back-up, etc.,

(g) Measures that violate the then current versions of the National Electric Code, the National Electric Safety Code, the Uniform Building Code or the Uniform Mechanical Code,

(h) Utility financed measures that do not result in savings in the base utility expenditures sufficient to cover the project costs.

GC.17.2 Facility Performance Requirements of ECMs. ECMs proposed by the Utility shall conform to the following facility performance standards:

(a) Lighting levels shall meet the minimum requirements of the then current Illuminating Engineering Society (IES) Lighting Handbook,

(b) Heating and cooling temperature levels shall meet Government design standards,

(c) ECMs shall permit flexible operation of energy systems for changes in occupancy levels and scheduling of facilities. In proposing an ECM, the Utility may assume the building function will remain constant unless otherwise indicated by the Government.

GC.18 Task Orders. Following the evaluation of the ECM proposal, the Government may elect to execute a T.O. with the Utility for the evaluation, implementation or operation and maintenance of the ECM. If requested by the Government, the Utility will provide or obtain financing on terms at least as good as those available to customers in a comparable service class, or with a comparable risk profile, considering the nature of the security interests to be granted, if any, and other conditions affecting the cost of financing.

The T.O. may have five phases; Audit (when applicable), Feasibility Study Phase, Engineering and Design Phase, Implementation Phase and Operation and Maintenance Phase. Because the extent of all the work is unlikely to be known at the time the T.O. is entered into, these phases shall be line items under the T.O., and shall be issued with an estimated Termination

Schedule at the time the T.O. is executed. However, work will not commence on a particular phase unless and until a *statement of work and a price for that phase have been agreed* upon.

Following completion and Acceptance of the Feasibility or Engineering and Design Phases, the Government may elect to (i) pay the ECM Cost for each completed Phase within thirty (30) calendar days of being invoiced, or (ii) defer payments for that Phase until the end of the next Phase at which time the Government shall pay the ECM Cost for each completed Phase within thirty (30) calendar days of invoice, or (iii) include such amounts in the ECM Cost, if the Government elects to proceed with the Implementation Phase. If the Government elects not to proceed with the next Phase, it shall pay the Utility the ECM Cost for the prior completed Phases, plus a Carrying Charge as negotiated by the parties in the T.O. A decision to proceed or not to proceed with the next Phase must be made within sixty (60) days of receipt of a written request from the Utility. Only the Contracting Officer shall be authorized to exercise the Government's option to proceed to the next Phase, and such exercise shall be provided in writing within sixty (60) days of receipt of a statement of work and price.

Government finance payments for the Implementation Phase shall begin on the date of the first Utility bill following the 30 day period after the Government takes possession of all or part of the ECM as provided in FAR, Part 36, Subpart 36.511, and a satisfactory ECM Performance Verification as defined in the T.O. and pursuant to Section GC.12 of this Agreement.

The timing and amount of Government payments of appropriated funds for the Operation and Maintenance Phase shall be determined in the T.O.

The T.O. shall be subject to any legally required Federal Acquisition Regulations. Because services may vary widely from one T.O. to another, the Contracting Officer will insure that the appropriate FAR clauses from the FAR matrix found at FAR, Part 52, Subpart 52.301, are incorporated into any contract entered into by the parties for services provided by the Utility under the T.O.

GC.19 ECM Feasibility Study Phase. The Task Order shall set forth a scope of work for a detailed study to determine whether particular ECMs proposed by the Utility are feasible (the "Feasibility Study"). The Task Order shall specify the terms for the completion of the Feasibility Study and establish a price for the Feasibility Study. The Government will pay the Utility the agreed-upon price for the Feasibility Study in accordance with the T.O. If the Government elects to proceed with the Engineering and Design Phase as set forth below in Paragraph GC.20, the cost of the Feasibility Study shall be rolled into the Engineering and Design Phase ECM Cost. The Feasibility Study will provide, at a minimum, the following information:

Technical Factors:
(a) Audits of energy consumption of existing equipment and facilities, including estimated energy and cost savings, and proposed retrofit costs and financial incentives/rebates,
(b) Water audits of supply and utilization facilities, if specified by the Government,
(c) Equipment to be removed or replaced, and new equipment to be installed,
(d) Specifications, including catalog cuts, for new equipment. Specifications should include (as applicable): power rating, estimated energy consumption, input/output, power ratio, lighting level and estimated equipment life,
(e) Operation and maintenance procedures required after ECM implementation (if significantly altered by the ECM),
(f) Training that will be provided for the proper operation and maintenance of ECPs, including details on how many hours of training will be provided and how many people will be trained,
(g) Electrical and mechanical sketches for all ECPs that involve changes to existing systems, (sketches will not be required for ECPs involving only component replacement),
(h) Government support (e.g. minor changes in Government operation, movement of equipment, etc.) required during implementation of the ECM,
(i) Utility interruptions needed for implementation of each ECP by type (gas, electricity, water, etc.), extent (room number, entire building, etc.) and duration,
(j) Identification of potential adverse environmental effects,
(k) Any documentation required to comply with applicable environmental laws,

(l)   The estimated construction time in calendar days, showing significant milestones,

(m)   The estimated annual energy savings in kilowatt-hour and kilowatt demand of electricity, decatherms of natural gas and cubic feet of water for the life of each ECP, including all assumptions and detailed calculations showing how savings were determined,

(n)   The estimated equipment life for each ECP,

(o)   A proposed method to verify energy savings at the time of ECM Acceptance which shall be subject to Government approval,

(p)   Documentation that each proposed ECP has been recommended and selected without regard to fuel source;

Cost Factors:

(q)   Estimated annual operation costs (e.g. increased use of alternate fuel sources, replacement filters) and increased maintenance costs (e.g. relamping with a higher cost product, etc.),

(r)   Total estimated ECM Cost to the Government,

(s)   Estimated breakdown of financial incentives/rebates for each ECM (if any) in a format mutually agreeable to the Parties,

(t)   Estimated Cost-of-Money Rate (percent),

(u)   Estimated annual energy and operation and maintenance cost savings including details on estimated annual savings for each area of savings, such as lighting, controls, motors and transformers,

(v)   Estimated breakdown of implementation costs for each area of energy savings, such as lighting, controls, motors and transformers,

(w)   Estimated costs for replacing existing components and installing new components/systems shall be listed separately,

(x)   Estimated unit costs for major components and systems,

(y)   Estimated Life Cycle Cost Analysis prepared in accordance with the then current edition of the <u>Energy Prices and Discount Factors for Life-Cycle-Cost Analysis</u>, published as the annual supplement to the National Institute of Standards and Technology (NIST) Handbook 135.

GC.20 ECM Engineering and Design Phase. After evaluation and Acceptance of the feasibility study, the Government may elect to proceed with the Engineering and Design Phase. Prior to proceeding, the Parties shall agree upon a statement of work for all engineering and design services necessary for the implementation of a particular ECM, a time frame for completion of the work, and a price or cost cap for engineering and design work for the ECM. If the Government elects to proceed with the Implementation Phase as set forth below, the cost of the engineering and design work shall be rolled into the total ECM Cost. This T.O. shall include an estimated amortization schedule for the ECM.

GC.20.1 Verification of Floor Plans. The Utility will verify the accuracy of any floor plans provided by the Government.

GC.20.2 Government Design Review. Task Orders shall permit adequate time for Government review of engineering and design work at 35% and 95% design completion, or at any other stage, as negotiated in the T.O.

GC.20.3 Site Plans. If proposed ECMs require installation outside existing buildings or structures, a site plan showing recommended siting of ECMs shall be prepared for Government review and approval. Site plans shall be submitted as part of the Utility's proposal. It is recommended that the Utility propose alternate sites for review in case the primary site is unavailable.

GC.20.4 ECM Implementation Proposal. Upon completion and Acceptance of the Engineering and Design Phase, the Utility will submit to the Government an ECM implementation proposal (the "Proposal"). If requested by the Contracting Officer, the Utility will be required to present a briefing to the Government explaining the Proposal. At a minimum, the Proposal shall include all pertinent technical and cost factors listed in Paragraph GC.19 of this Agreement plus a copy of subcontractor(s) bid(s). The Proposal shall also set forth negotiated pricing criteria that describes the method for determining the prices to be paid to the Utility for supplies or services. The Government shall evaluate the Proposal for technical soundness and price reasonableness. If the Government elects to proceed with the ECM, the Utility and Government shall agree upon a complete scope of work with specifications, time for performance, ECM Cost, source and cost of capital or financing, payment terms, amortization schedule and final Termination Schedule. If the Contracting Officer deems it appropriate, the Utility will provide acceptable performance and payment bonds.

GC.21 ECM Implementation Phase. The Utility shall perform work in accordance with the T.O. The following provisions shall apply to ECM implementation work performed pursuant to T.O.s executed under this Agreement, unless exceptions are provided in the T.O.

GC.21.1 Pre-Work Requirements. Prior to commencing ECM implementation Work on a T.O., the Utility shall meet with the Contracting Officer or COR at a time mutually agreeable to the Utility and the Contracting Officer, to discuss and develop mutual understandings relative to safety, scheduling, performance, obtaining necessary permits, and administration of the Implementation Phase. Prior to commencement of on-site work, written approval of the following shall be obtained from the Contracting Officer by the Utility:

(a)   Utility's proposed implementation schedule indicating the installation period and time required for delivery of equipment,
(b)   Evidence that the required insurance has been obtained.

GC.21.2 Interruptions. The Utility shall arrange on-site work to minimize interference with normal Government operation. All interruptions shall be made outside occupied periods whenever possible and coordinated with the Contracting Officer or COR. The Utility shall endeavor to keep the duration of utility interruptions to a minimum. Requests for utility outages shall be submitted for approval, in writing, as specified in the T.O. The request shall include the approximate duration, date, time and reason for the interruption. Utility interruptions include, but are not necessarily limited to, the following systems:

(a) Electrical,
(b) Natural Gas,
(c) Sewer,
(d) Steam,
(e) Water,
(f) Telephone,
(g) Computer cables.

GC.21.3 Construction Documentation. The Utility shall provide construction drawings and specifications, certified by a registered engineer or architect, as applicable, to ensure compliance with all applicable federal, state and local codes and regulations as required by individual T.O.s.

GC.21.4 Standardization of Materials. All materials proposed to be installed pursuant to this Agreement shall be readily, commercially available, and as similar in form, fit and function to each other as is practicable to allow efficient provisioning of replacement parts.

GC.21.5 Water Conservation Measures. The Utility will consider water conservation in all ECMs. The Utility will obtain rebates from the local water utility if available. Rebates, if any, shall be applied to the cost of the project.

GC.21.6 Operation and Maintenance Manuals. At the time of Government Acceptance of a completed ECM, the Utility shall furnish, for the equipment specified, operation and maintenance manuals and recommended spare parts lists identifying components adequate for competitive supply procurement for operation and maintenance of ECM equipment. The operation and maintenance manuals shall include maintenance schedules for all equipment. The scope of each manual shall be agreed upon in the T.O.

GC.21.7 Government Personnel Training for ECPs. The Utility shall train Government personnel, as required, to operate, maintain, and repair ECM equipment and systems. The date and time of training shall normally be coordinated with the Contracting Officer or COR prior to Acceptance of the ECM. The cost for such training shall be included in the ECM Cost.

GC.21.8 As-Built Drawings. Within forty-five (45) calendar days after Government Acceptance of each installed ECM, the Utility shall submit as-built drawings to the Contracting Officer or COR. Drawings will not be required for component replacement. Drawings shall include at a minimum:

(a) The installation (i.e., form, fit, and attachment details) of the interface between ECM equipment and existing Government equipment,
(b) The location and rating of installed equipment on building floor plans.

GC.21.9 Installation. The Utility will arrange for the installation of approved ECMs and construction oversight and verify that the designed and specified energy efficiency equipment and/or system modifications are properly supplied or installed in a manner that will give the intended long term demand and energy reductions. The Utility will select Subcontractors in accordance with Paragraph GC.7 above.

GC.22 Operation and Maintenance Phase. The Government may elect to have the Utility perform the operation and maintenance on part or all of the ECM. Before exercising its option for this Phase, the Government and Utility shall agree upon a complete scope of work with specifications, schedules, warranties and cost.

GC.23 Required FAR Clauses. The following FAR clauses are required to be included in any contract with the Government:

52.203-3 Gratuities,
52.203-5 Covenant Against Contingent Fees,
52.203-7 Anti-Kickback Procedures,
52.222-3 Convict Labor,
52.222-25 Affirmative Action Compliance,
52.222-26 Equal Opportunity,
52.223-6 Drug Free Workplace,
52.233-1 Disputes.

WARRANTIES AND REMEDIES

WR.1 Warranties. The Utility shall pass through to the Government all warranties on equipment installed pursuant to a T.O. In addition, the Utility will provide, from the date of Acceptance or Government Possession of an ECP, whichever is earlier, a one year comprehensive wraparound warranty guaranteeing that the equipment installed shall perform in accordance with the specifications agreed upon between Government and Utility, as set forth in the applicable T.O.

In the event the Utility provides O&M services, a separate warranty will be negotiated for such services, in accordance with FAR Part 52, Subpart 52.246-20.

WR.2 No Other Warranties. The warranties set forth in WR.1 are exclusive and in lieu of all other warranties. The Utility makes no other representations or warranties of any kind with respect to the services and products it provides pursuant to this Agreement and subsequent T.O.s, The Utility does not guarantee any level of energy or water savings or cost reductions.

WR.3 Utility Limitation of Liability. The Utility shall not be liable for any special, incidental, indirect, or consequential damages, connected with or resulting from the performance or non-performance of work under this Agreement or subsequent T.O.s. In addition, the Utility shall not be liable under its warranty to the extent that damages are caused by Government negligence.

WR.4 Utility Default. The Government and Utility agree that Utility default provisions will be governed by those FAR clauses applicable to specific circumstances. A determination of applicable FAR default clauses will be made by the Contracting Officer for a specific T.O.

WR.5 Prompt Payment. As required in FAR, Part 32, Subpart 32.903, the Government shall promptly pay ECM utility bills. Late payments shall accrue interest as provided in FAR, Part 32, Subpart 32.907.

WR.6 Disputes. Disputes that arise under this Agreement and subsequent T.O.s shall be governed by the applicable dispute provisions found at FAR, Part 33, Subpart 33.2.

WR.7 Differing Site Conditions. In the event site conditions differ materially from those contained in the T.O. additional costs incurred by the Utility due to the differing conditions shall be negotiated prior to work, and the ECM Cost shall be increased to reflect an equitable adjustment as permitted in FAR, Part 36, Subpart 36.502.

WR.8 Suspension of Work. In the event Work is delayed, suspended or stopped by the Government, FAR, Part 42, Subpart 42.13 shall apply.

FINANCING AND PAYMENT PROVISIONS

FP.1 Energy Savings and Financing: It is intended that the life-cycle energy and related savings achieved from the implementation of an ECM funded or financed in a UESC project will produce financial savings to the

Government that are equal to or greater than the cost of implementing the ECM, including the cost of financing, if applicable, provided under this Agreement. The payment term shall be in accordance with agency policy following current legislation, legal opinions, and agency guidance.

FP.2 Financial Incentives, Rebates, and Design Assistance: The Utility will provide to the Government the same financial incentives, rebates, design review, goods, services, and/or any other assistance provided without charge, that is generally available to customers of a similar rate class or size. Incentives that may be available are to be identified in the preliminary audit report provided according to Paragraph GC.15 and the ECM implementation proposal provided according to Paragraph GC.20.4.

If rebates are available and have been applied for by the government and such funds have been set aside, then the Utility shall provide a separate Letter of agreement clarifying timelines and responsibilities of both parties and guaranteeing rebates and other incentives from the Utility to the Government.

The Utility shall also be responsible for determining the source, value, and availability of any applicable financial incentives to the project offered by the state and others in which the facility is located, and if the value of the incentives exceeds the administrative costs to be incurred by the Utility or the Government in acquiring such incentives.

The Utility shall be responsible for coordinating with the Agency Contracting Officer as to the preparation of any and all documentation required to apply for any such applicable financial incentives and to effectively apply such incentives to the capital cost of the project.

Rebate disbursement options:

Option 1: Utility shall apply rebate to the next payment due to reduce capital cost of the project

Option 2: Where allowable by the Public Utility Commission, Government may assign rebate to a third party to reduce the construction costs and thereby reducing the total amount financed.

Option 3: Rebate may be accepted as a credit on the utility bill

FP.3 Calculation of Payment. Payment for accepted ECMs shall be equal to the ECM Cost amortized over a negotiated term. In accordance with 10 U.S.C. Section 2865, the cost of financing, if any, for any completed ECM shall be recovered under terms and conditions no less favorable than those for others in the same customer class. Monthly payments will commence on the date of the first Utility bill following the 30 day period after the date the Government takes Possession of the ECM and ECM Performance Verification Testing, as required by GC.12 and negotiated in the T.O., is satisfactorily completed.

FP.4 Buydown. The Government reserves the right; at any time following Acceptance, but prior to final payment, to buydown the outstanding T.O. payments without penalty by giving thirty (30) days written notice to the Utility. Upon such buydown, the Government shall pay to the Utility a negotiated amount to include an additional finance charge based on an indexed formula, which reduces the financiers risk and reduces the cost of buydown to the agency, or a termination schedule. Monthly payments will continue at the same level but the term of ECM financing will be shortened to reflect the amount of the buydown payments.

FP.5 Pre-Acceptance Buyout. In the event the Government desires to terminate a Task Order for any reason (including, without limitation, for convenience) prior to Acceptance, the Government may do so by giving written notice to the Utility thirty (30) days prior to the effective date of such termination. The Government shall pay to the Utility a negotiated amount to include an additional finance charge based on an indexed formula, which reduces the financiers risk and reduces the cost of buyout to the agency, or a termination schedule which will be described in Attachment A of the Task Order. If a termination occurs for the convenience of the Government, the amount payable pursuant to this paragraph shall be deemed as an allowable cost under FAR. (See Part 17 and Part 52, Subpart 52.249-2.)

FP.6 Post-Acceptance Buyout. In the event the Government desires to terminate a Task Order for any reason (including, without limitation, for convenience) after Acceptance, the Government may do so by giving written notice to the Utility thirty (30) days prior to the effective date of such termination. The Government shall pay to the Utility a negotiated amount to include an additional finance charge based on an indexed formula, which reduces the financiers risk and reduces the cost of buyout to the agency, or a termination schedule which will be described in Attachment B of the Task Order. If a termination occurs for the convenience of the Government, the amount payable pursuant to this paragraph shall be deemed as an allowable cost under FAR. (See Part 17 and Part 52, Subpart 52.249-2.)

FP.7 Assignment of Claims. Government payments under each T.O. executed pursuant to this Agreement may be assigned pursuant to FAR, Part 52, Subpart 52.232.23 "Assignment of Claims." Any bank, trust company or other financing institution that participates in financing an ECM shall not be considered a Subcontractor of the Utility. Any "Assignment of Claims" must comply with the provisions of FAR, Part 32, Subpart 32.8.

FP.8 Novation. The Parties agree that if, subsequent to the execution of this Agreement, it should become necessary, or desirable, to execute a "Novation Agreement," said Novation Agreement will comply with the provisions of FAR, Part 42, Subpart 42.12 and will be in the form as provided at FAR, Part 42, Subpart 42.1204.

SPECIAL REQUIREMENTS

SR.1 Environmental Protection. The Utility shall comply with all applicable federal, state and local laws, regulations and standards regarding environmental protection ("Environmental Laws"). All environmental protection matters shall be coordinated with the Contracting Officer or designated representative. The Utility shall immediately notify the Contracting Officer of, and immediately clean up, in accordance with all federal, state and local laws and regulations, all oil spills, hazardous wastes, (as defined at 42 U.S.C. §9601), and hazardous materials (as defined at 49 C.F.R. Pt. 172) collectively referred to as "Hazardous Materials," resulting from its operation on Government property in connection with the implementation of ECMs. The Utility shall comply with the instructions of the Government with respect to avoidance of conditions that create a nuisance or create conditions that may be hazardous to the health of military or civilian personnel.

SR.2 Environmental Permits. Unless otherwise specified, the Utility shall provide, at its expense, all required environmental permits and/or permit applications necessary to comply with all applicable federal, state and local requirements prior to implementing any ECM in the performance of a T.O. executed pursuant to this Agreement. If any such permit or permit application requires the signature or other cooperation of the Government as owner/operator of the property, the Government agrees to cooperate with the Utility in obtaining the necessary permit or permit application.

SR.3 Handling and Disposal of Hazardous Materials. Not withstanding the provisions of the FAR, Part 52, Subparts 52.236-2 "Differing Site Conditions" and 52.236-3 "Site Investigations and Conditions Affecting Work," the Government understands and agrees that (i) the Utility has not inspected, and will not inspect, the project site in connection with a proposed ECM for the purpose of detecting the presence of pre-existing Hazardous Materials that relate to an ECM or any project site, and (ii) the Government shall retain sole responsibility for the proper identification, removal, transport and disposal of any fixtures, components thereof, or other equipment or substances incidentally containing pre-existing Hazardous Materials, except as specifically agreed to by the Utility pursuant to Paragraphs SR.4 and SR.5 (below).

If the Utility, during performance of the work under a T.O. executed pursuant to this Agreement, has reason to believe that it has encountered or detected the presence of pre-existing Hazardous Materials, the Utility shall stop work and shall notify the Government. The Government will evaluate the site conditions and notify the contractor of the results of this evaluation. The Utility shall not be required to recommence work until this situation has been resolved. Any delay resulting there from shall be grounds to request an increase in the ECM Cost to the extent that such delay increases ECM Costs.

SR.4 Asbestos and Lead-Based Paint. To the extent provided for in a T.O. executed pursuant to this Agreement, in connection with the implementation of any ECM, the Utility may agree to remove pre-existing asbestos containing material or lead-based paint, incidental to implementation of an ECM. However, unless the Utility explicitly agrees in said T.O. to perform any portion of the testing, removal or abatement of the pre-existing asbestos or lead-based paint as part of the scope of work for any ECM, and unless the T.O. specifically references this Paragraph SR.4, the Government shall be deemed to be solely responsible as provided for in Paragraph SR.3.

If the Utility in the course of ECM implementation disturbs suspected lead-based paint or asbestos containing material, the Utility may propose to the Government that the Utility will perform any portion of the testing, removal, or abatement of the lead-based paint or asbestos containing material. Said proposal will include the requested increase in the ECM Cost on account of such additional work. The Utility will not commence work involving additional cost without approval of the Contracting Officer. In the absence of an agreement to the contrary, the provisions of Paragraph SR.3. (above) shall apply.

In the event the Utility agrees to include any portion of the testing, removal or abatement of the asbestos within the scope of work for an ECM implemented as described above in this Paragraph, the hazardous waste manifests or other shipping papers shall identify the Government as the sole generator of the Hazardous Materials.

SR.5 Refrigerants, Fluorescent Tubes and Ballasts. To the extent provided for in a T.O. executed pursuant to this Agreement in connection with the implementation of any ECM, the Utility shall remove and/or dispose of all ozone depleting refrigerants, fluorescent tubes and fluorescent magnetic core and coil ballasts incidental to an ECM to the Hazardous Materials Disposal site (HAZMAT) on the installation. If there is no HAZMAT on the installation, the above Hazardous Materials will be disposed in accordance with all applicable federal, state and local laws and regulations, provided however, that the hazardous waste manifests or other shipping papers shall identify the Government as the sole generator of the Hazardous Materials.

Appendix F

DOE/EEI Model
Agreement Explanation

INTRODUCTION

The attached document serves as a model for the development of formal agreements between a Federal civilian Agency and its serving Utility for the procurement of energy services on a "designated" or "sole" source basis.

The Energy Policy Act of 1992 ("EPAct") establishes as a federal government goal, identifying and implementing all energy and water conservation projects with a payback of ten years or less. Executive Order 13123, signed on June 3, 1999, requires all federal agencies to achieve a 30% reduction in facilities energy use and a 20% improvement in industrial energy efficiency by the year 2005, relative to 1990 levels. The Order requires a 35% reduction in energy use at facilities and a 25% improvement in industrial energy efficiency by 2010, relative to 1990 levels.

Accomplishing these goals will result in a billion dollars of annual savings in energy and water cost and an equivalent savings in maintenance costs for the Federal government. In addition, energy and water efficiency improvements will improve the installations' infrastructure, readiness, personnel quality of life and productivity, and reduce the environmental impact of Federal facilities.

The benefits of the Executive Order goals can be realized even though internal Federal technical and financial resources are being reduced. EPAct and Executive Order 13123 allow government facilities to obtain energy services initially paid for by the private sector and repaid by the Federal government from energy and water bill savings. These energy services may be purchased from gas and electric   Utilities on a "designated source" or "sole source" basis.

The legal authority for such "sole source" acquisitions comes from 42 USC 8256. On _____, 1998, representatives of the Department of Energy, General Services Administration, _____, ... who are involved in utility acquisition policy, legal and regulatory matters met to review the authority contained in 42 USC 8256 and its impact on the legal basis for the non-competitive procurement of energy demand management and energy conservation services from gas and electric Utilities. Those attending concurred in the following position statement and agreed to apply the statement in exercising their acquisition responsibilities within each Agency:

"Contracting officers of a Federal Agency may procure on a sole source basis from gas or electric Utilities (however, not from unregulated subsidiaries of such Utilities) the design and implementation of cost-effective demand and conservation incentive programs and services including but not limited to the following:

1.   *Energy & water conservation measures including audits and surveys, design and construction, and the operation and maintenance of systems provided.*

2.   *Energy Management Services including services related to DSM, Incentive Programs, Metering, and Energy Management Control Monitoring Systems.*

3.   *The operation and maintenance of existing energy and water systems and equipment including utility distribution and collection systems, generation and treatment systems, energy related equipment, systems, and facilities for buildings and metering.*

4.   *Financial assistance*

5.   *Training*

42 USC 8256 should be cited as the legal authority for such sole source acquisition. Any programs or services obtained shall be limited to those with a positive net present value of 10 years or less.

Clearly, 42 USC 8256 allows a departure from some traditional contracting practices. It gives Contracting Officers the ability to sole source energy service improvements to its utility company. While 42 USC 8256 provides the authority for Federal agencies to contract directly with the Utility in order for "sole sourcing" to occur, the Utility is free to hire Subcontractors, including its subsidiaries, to perform the work so long as the Utility can prove the cost reasonableness of the work to the Contracting Officer.

While use of the Civilian Model Agreement is broadly recommended, there is no requirement that Utilities or the agencies use all or part of it when entering into their own conservation/demand side management agreements.

The Civilian Model Agreement is structured using a task order format under which specific Task/Delivery Orders can be written. It includes recommended language for certain terms and conditions that the agencies and representative utilities have already endorsed.

The Civilian Model Agreement was drafted to provide language that had sufficient specificity to be useful but was not so narrow that it would inadvertently limit the parties' ability to develop the broad variety of projects that will arise. For this reason, many of the Agreement's sections defer decisions to the Contracting Officer and Utility representative with the understanding that the decisions will be resolved during the negotiation of individual Task/Delivery Orders or in some cases, the negotiation of agreements patterned after the Civilian Model Agreement.

It is important to note that the Civilian Model Agreement has been structured such that there are several ways it can be implemented. The Civilian Model Agreement may be executed as a stand alone agreement, an attachment to a GSA Areawide Utilities Contract Exhibit, or as a modification to an existing utility service contract.

The Civilian Model Agreement and this Explanation reflect the experience of many active and successful Utilities and agencies in the area of negotiated energy services agreements. It incorporates the lessons learned by both the Utilities and the Agencies in implementing tens of millions of dollars of energy projects. This is not to say that the Civilian Model Agreement is perfect, only that it includes and reflects the lessons learned over the past five years.

The Explanation discusses in detail the reasons behind the specific terms and conditions of the Civilian Model Agreement. It mirrors the structure of the Civilian Model Agreement and intends to provide the reader with an insight as to why a particular structural approach was taken and why certain language was included. All capitalized words are defined terms in the Civilian Model Agreement.

THE MODEL AGREEMENT EXPLANATION

GC.1 Purpose.—Self Explanatory

GC.2 Definitions.
Acceptance—Acceptance can take place at the end of each Phase and at the final completion of the ECM. Acceptance is used to determine when

title is transferred and when the Utility's warranty begins. Payment, however, is not triggered by Acceptance of an ECM. As explained in FP.3, payment begins only after the Government takes Possession of an ECM and the Utility successfully completes Performance Verification Testing, both of which can occur prior to Acceptance of a specific Phase.

Agency—Self-Explanatory.

Carrying Charge—The term Carrying Charge refers to the rate at which the Utility accrues interest on money advanced for the Feasibility, Engineering and Design and Implementation Phases.

Contracting Officer—Self Explanatory

Contracting Officer's Representative (COR) or Contracting Officer's

Technical Representative (COTR)—Self Explanatory

Task Order (T.O.)—Task Orders (sometimes called Delivery Orders) are the mechanisms through which specific projects (ECMs) are negotiated and implemented.

   If a Federal facility and the local Utility choose to enter an agreement patterned after the Civilian Model Agreement, they will need to write specific Task Orders or Task Orders for each ECM. These Task/Delivery Orders may include additional terms and conditions applicable only to that specific project.

Energy Conservation Measure (ECM)—ECMs consist of one or more Energy Conservation Projects (ECPs) dealing with a broad range of energy needs. An ECM should have a 10 year payback or less to conform with the utility services contract ten year term limitation per FAR} Part 41. Should this 10 year term be altered in subsequent legislation, and/or regulation, the new term should be substituted for the present 10 year requirement only for ECMs entered into after the effective date of the legislation.

Energy Conservation Measure Cost (ECM Cost)—the ECM Cost is the total project cost and consists of three major components: 1) Work (direct costs), 2) finance charges, and 3) authorized overhead, carrying cost, taxes and profit. It is this amount that is amortized to determine the Government's monthly payment if an extended payment option is desired.

Energy Conservation Project (ECP)—The ECP is a specific energy or water project. A detailed list of ECP examples is located in GC.17.

Government—Self-Explanatory.

Occupied Period—Self Explanatory

Quality Assurance Evaluator (QAE)—The Government personnel responsible for monitoring completion of the ECM.

Quality Control—The process used by the Utility to ensure the ECM is correctly implemented.

Possession—Possession triggers when the warranty period begins and is an element in determining when monthly Government payments begin. While the phrase "beneficial occupancy" is given as the definition of Possession, the word "Possession" is used because it is the term used in the warranty language of the FARs. Within the context of the Civilian Model Agreement, the word Possession and the phrase "beneficial occupancy" have the same meaning.

Subcontractor—Self Explanatory

Termination Schedule—The Termination Schedule and how it is referenced within a Task/Delivery Order is critical to the interest rate the Government will be required to pay for financing. Financiers require some level of certainty such that they know how much they will be paid in the event the Task/Delivery Order is terminated. If the Government has the ability to terminate the Delivery/Task Order and it is unclear what it will have to pay to do so, financiers may charge a higher interest rate or refuse to finance an ECM altogether. Therefore, it is essential to have an agreement and Task/Delivery Orders that make clear what the Government will have to pay at any point in time should the Government decide to terminate the Task/Delivery Order.

In the case of termination during construction financing, the parties should agree upon a termination formula (see Section FP.5) due to the difficulty in predicting the amount of construction dollars that will have been spent at any point in time.

Utility—Self-Explanatory.

Work—Self Explanatory

GC.3 Term—This Civilian Model Agreement may be terminated with 90 days notice by either party. Any Task Orders or other agreements entered into under this Agreement shall remain in full force and effect even after the Civilian Model Agreement is terminated. To help provide certainty to the financiers, it is important to expressly state that previous obligations will not be altered by the termination of the Civilian Model Agreement.

There is no prescribed contract term in the Civilian Model Agreement, however the parties will need to establish such a date if they choose to pattern their agreement after the Civilian Model Agreement. The parties should also decide how, if at all, their agreement will be influenced by the termination of higher level contracts such as an Areawide Contract.

GC.4 Services to be Provided by the Utility—The Utility and its subsidiaries may provide a broad range of services (described in GC.17) under the Civilian Model Agreement. Any Work performed by either the Utility or its Subcontractors must be "price reasonable" as determined by the Contracting Officer relying on FAR 15.8 for guidance.

The services to be provided for a specific ECM will generally be in Phases, including a Preliminary Audit Phase, Feasibility Study Phase, Design and Engineering Phase, Implementation Phase and Operations and Maintenance Phase. The Government may withdraw from its agreement with the Utility at any point in this staged process. Furthermore, the Government may order any Phase of services without being obligated to order other Phases. For example, the Government may have the Utility perform the Engineering and Design of an ECM and have another Contractor implement the ECM. Conversely, the Government may have an ECM designed by another firm and implemented by the Utility.

GC.5 Information—Self Explanatory

GC.6 Relationship of Parties—This Section makes it clear that the Utility and its Subcontractors are independent Contractors and are not considered employees or agents of the Government. In addition, this Section explicitly requires the Utility to ensure that Subcontractor guarantees and warranties flow to the Government.

GC.7 Subcontractor Selection—The selection process described in this Section is meant to summarize the FAR requirements for selecting Subcontractors to perform Work. The Parties should look to FAR Subpart 15.6 for guidance in selecting Subcontractors.

GC.8 Authority of Contracting Officer—Self Explanatory

GC.9 Ownership of Work Product—This Section makes it clear that the Government owns the work product at the end of any Phase of this process and is free to use that work product in any manner it chooses. One exception to this general rule may be software systems and other intellectual properties which may be proprietary to the Utility or its suppliers.

More importantly, this Section also deals with when ownership/title to Work transfers to the Government. Section (d)(4)(C) of 10 USC 2865 states, "Such title may vest at such time during the term of the agreement, or upon expiration of the agreement, as determined to be in the best interests of the United States." While this statute applies only to the Department of Defense, the logic of the intent of Congress was believed to apply to other federal agencies. The Civilian Model Agreement recommends that title transfer upon Acceptance because this is often the least expensive alternative and it does not diminish the Government's ability to seek relief should the Utility improperly perform its duties under the Task Order.

The reason it costs less to take title at Acceptance is due to the treatment of the associated taxes and insurance. If the Government does not take title at the time of Acceptance, it resides with the financier which means the Utility will be responsible for paying taxes and insurance. The Utility's cost of insuring the Work and paying any taxes will be passed on to the Government thereby increasing the cost of the ECM. If the Government takes title, it will self-insure and may legally avoid personal property taxes which can represent significant savings.

There has been some concern about taking title at the time of Acceptance in the belief that if the Utility/financier retains title it gives the Government more leverage if something goes wrong. This is not the case.

In the first place, the Government does not start paying the Utility until after the project is accepted. To the extent the Government is dissatisfied with the project implementation, it has all of the remedies under the FARs and has the ultimate leverage because the Utility is paying all of the implementation costs and the Government does not have an obligation to pay until it verifies that the ECM works as designed. Once the project is complete and permanent financing is in place, the Government has an obligation to pay a fixed monthly amount. The Government's remedies at that time are governed by the FARs, just as they would be if title remained with the financier.

GC.10 Responsibility for Operations and Maintenance—This Section assumes that the operation and maintenance of equipment installed pursuant to the Agreement will lie with the Utility. However, utility O&M is simply an option and not a requirement under the Civilian Model Agreement.

GC.11 Government Projects—The Civilian Model Agreement is not intended to restrict the Government from implementing other energy projects independent of the Work the Utility will perform. Although such Work will need to coordinated, the Government is free to pursue other energy projects on its own or with other companies.

GC.12 ECM Performance Verification—Performance verification provides the Government with assurance that the ECM will perform as designed. Each ECM should have its own measurement and verification (M&V) plan that takes into account the cost of the plan, complexity of the project and risk sharing between the Government and Utility. The measurement and verification plan anticipated in the Civilian Model Agreement would be a plan similar to Option A of the Department of Energy M&V Guidelines for Federal Energy Projects (DOE/GO-10096-248-February 1996). This alternative requires confirmation that the equipment is installed per specifications, is operating and meets all functional tests at the time of Acceptance. It does not assume ongoing M&V with subsequent adjustments to Government payments.

GC.13 Emission Credits—Self Explanatory

GC.14 Order of Precedence—This Section was included in the Civilian Model Agreement to draw attention to the need to ensure consistency or determine an order of precedence among all of the agreements that make up a relationship between a Utility and the Government. For example, what happens if the Civilian Model Agreement is written as an exhibit to an Area Wide Contract ("AWC") and the AWC is terminated? Can the Government still continue to execute T.O.s under the Civilian Model Agreement? Many of the Drafters believed that the Task/Delivery Order should take precedence over the Civilian Model Agreement which in turn should take precedence over a higher level agreement such as an Area Wide Contract. Others were concerned that such an order of precedence might be contrary to the order of precedence found in most Government contracts and would create confusion. Due to the variety of opinions on this issue, the Civilian Model Agreement leaves it to the Utility and Contracting Officer to determine what order of precedence makes the most sense. However, this Section is included as a reminder to the negotiating parties that they need to discuss and resolve the issue of what happens if there are inconsistencies among agreements.

GC.15 Preliminary Audits—The Preliminary Audit is often the first step a Utility and the Government take in identifying economical energy projects. The Preliminary Audit is frequently provided at no cost to the Government. The Civilian Model Agreement suggests types of information the Government should expect in this phase of project identification. The parties should feel free to add to or delete from this list. The Utility may request that the Government keep the audit report confidential. While the ideas suggested in the audit are not confidential, the audit report itself may be considered confidential, if such treatment is required of other customers.

GC.16 ECM Proposal—After the Government has received the results of the Preliminary Audit, it may ask the Utility to submit a proposal detailing energy improvements available to the Government. The proposal will be prioritized and contain a number of key elements such as the cost of the project and estimated savings. Should the Government decide to proceed, the Utility and the Government will move to the Feasibility Study Phase where the ECM proposal will be further refined.

GC.17 Energy Conservation Projects (ECPs)—The list of ECPs provided in Section GC.17 reflects the range of services that Utilities can provide on a "designated" or "sole" source basis pursuant to 42 USC 8256. Section GC.17 (ss) is included to demonstrate that the types of projects listed in this Section are only examples and are not meant to be a complete list.

GC.17.1 ECM Restrictions—While 42 USC 8256 authorizes a broad range of energy services that can be provided by Utilities, there are some restrictions on those services that should be noted. Except in the cases of GC.17.1(c) and (f), all of the restrictions are unconditional. In (c) and (f), dealing with measures that result in increased water consumption and the use of electrical capacity reserved for other uses, the Civilian Model Agreement provides the Contracting Officer the flexibility to decide whether the increased energy savings are great enough to justify the use of more water or reserved capacity. For example, a Contracting Officer may elect to install an air conditioning evaporation unit which would use more water but save substantially more money on energy costs.

GC.17.2 Facility Performance Requirements of ECMs—Self Explanatory

GC.18 Task Orders—This Section describes the mechanics of how a Utility / Government relationship will work under the Civilian Model Agreement. The Civilian Model Agreement provides a format for Delivery / Tasking Orders that are issued for each ECM for a Government facility. Delivery / Task Orders will be issued for each ECM.

A Task Order can have up to five Phases—Audit (when applicable), Feasibility Study Phase, Engineering and Design Phase, Implementation Phase and Operations and Maintenance Phase. For each T.O. Phase, the Utility will provide a cost estimate and preliminary scope of services. The Government may proceed with a Phase of the T.O. only after it has received a complete scope of work and a price for that Phase from the Utility. For example, a Federal facility may sign the T.O. and commit to move ahead with the Feasibility Phase but it is not obligated to do anything more than complete the Feasibility Study.

If the Government decides not to proceed to the next Phase, it owes the Utility the cost of the Audit, if applicable, or if not, the Feasibility Study and interest (Carrying Charge) on money used to conduct the audit or study. If it decides to move to the next Phase, the cost of the

Feasibility Study and its Carrying Charge may be rolled into the cost of the next Phase. Should the Government decide to implement the project, the costs for earlier Phases will be included in the ECM Cost financing.

With respect to the O&M Phase, the Civilian Model Agreement recommends that O&M services and their scope, term, warranty and payment, be negotiated as a Phase separate from the Implementation Phase. Generally, O&M services will not be financed but will be paid out of savings generated by the project on an ongoing basis.

The last paragraph of Section 18 reflects an important part of the Civilian Model Agreement. Due to the extremely wide range of potential projects, services and combinations thereof that could be pursued under the Civilian Model Agreement, it is impossible to identify the FAR clauses, other than those listed in Section 23, that may apply to a specific Task Order. Therefore, in an effort to make the contracting process as streamlined and flexible as possible, the Civilian Model Agreement assumes that the Contracting Officers will determine what FAR provisions should be included in the Delivery / Task Order.

GC.19 ECM Feasibility Study Phase—The Feasibility Study Phase is the first step in refining the Preliminary Audit. In order to help the parties conduct a Feasibility Study that provides the Government with adequate information to decide whether to proceed to the Engineering and Design Phase, the Civilian Model Agreement provides a list of recommended technical and cost factors that should be addressed in the Feasibility Study Phase.

GC.20 ECM Engineering and Design Phase—After the Feasibility Study Phase, the ECM is further refined in the Engineering and Design Phase. At the end of this Phase, the Utility will provide the Government with a project proposal that includes financing costs and monthly payment and savings figures.

GC.20.1 Verification of Floor Plans—The Civilian Model Agreement recommends that the Utility, with the cooperation of the Government, be responsible for verification of floor plans. By assigning responsibility to the Utility, the parties will avoid later misunderstandings about verification responsibility if the project design and floor plans are inconsistent.

GC.20.2 Government Design Review—The Civilian Model Agreement provides for Government review of the project design at least twice during the engineering and design process. It suggests, at a minimum, reviews at 35% and 95% completion and encourages the parties to decide what design reviews make the most sense for their specific ECM.

GC.20.3 Site Plans—As was the case with verification of floor plans, the Utility has responsibility for reviewing site plans if the ECM installation is outside existing buildings. It is recommended that the Utility offer a preferred site plan and at least one alternative.

GC.20.4 ECM Implementation Proposal—At the end of the Engineering and Design Phase, the Utility is required to present the Government with a detailed project proposal. The Government will need enough information from this proposal to decide whether to proceed with implementation. This Section sets forth some of the recommended elements of the proposal. This information will permit the Government to determine for the ECM such issues as technical soundness, time for performance, price reasonableness, project cost and savings and a payment and termination schedule. It is at this point that the Government will decide whether to implement all or part of the proposed ECM.

GC.21 ECM Implementation Phase—Sections GC.21.1-GC.21-9 provide language regarding key areas the Government and Utility may want to address prior to beginning the implementation phase of the ECM.

GC.21.1 Pre-Work Requirements—The first step of the Implementation Phase is for the Government and Utility to agree on how the project will proceed with respect to issues of safety, scheduling, performance, obtaining necessary permits, and administration of the Implementation Phase. It is assumed that during this step, the parties will establish a schedule and protocol to insure that communications are effective throughout the ECM implementation process. In addition, it is recommended that the project not go forward until the Government has reviewed and approved the Utility's implementation schedule and has obtained evidence of all required insurance.

GC.21.2 Interruptions—Self Explanatory

GC.21.3 Construction Documentation—Self Explanatory

GC.21.4 Standardization of Materials—Self Explanatory

GC.21.5 Water Conservation Measures—This Section reinforces that ECMs which save water as well as energy should be pursued as authorized under 42 USC 8256. It also makes it clear that it is the Utility's responsibility to acquire local water company rebates for water conservation and credit them to the ECM Cost.

GC.21.6 Operation and Maintenance Manuals—Self Explanatory

GC.21.7 Government Personnel Training for ECPs—The Civilian Model Agreement recommends leaving the time and date of training up to the Contracting Officer and Utility. This timing will provide the flexibility to schedule the training when it is appropriate to the specific ECM. For example, if the Utility was going to perform all of the O&M on the equipment, the need to train Government personnel would be vastly different than if the Government was performing the O&M.

GC.21.8 As-Built Drawings—Self Explanatory

GC.21.9 Installation and Acceptance—This Section reinforces the notion that the Utility is the party ultimately responsible for the proper design and installation of the ECM.

GC.22 Operation and Maintenance Phase—This Section emphasizes that the O&M Phase may be separate and distinct from the Implementation Phase and that it may be an independent agreement between the Government and Utility with its own scope, terms, cost, payment and warranty.

GC.23 Required FAR Clauses—The FAR Clauses listed in GC.23 are required by law to be included in all federal government contracts and cannot be negotiated away by a Contracting Officer. They are listed to inform the Parties (especially those not familiar with Government contracting) of provisions that must be included in the agreement.

WARRANTIES AND REMEDIES

WR.1 Warranties—The Warranty Section of the Civilian Model Agreement once again reflects the philosophy that the Government should have to go to only one place (the Utility) to resolve contract performance issues. The Government should not need to pursue Subcontractors in the event an ECM is not working properly nor to be involved with determining relative "blame" if several Subcontractors are involved in an ECM performance problem during the term of the Utility's warranty.

WR.2 No Other Warranties—This Section makes it clear that the Utility is not offering warranties in addition to the wrap around warranty and the pass through of Subcontractor warranties.

This Section also makes it clear that the Utility is not necessarily guaranteeing energy or water savings beyond the Acceptance and verification period. The Utility will guarantee that the ECM will operate as designed at the time the Government takes Possession and successfully concludes monitoring and verification procedures. While some Utilities may be interested in guaranteeing savings during the term of the Task Order, most will not for two reasons. First, ongoing performance guarantees add significant financing and operational costs to the ECM. In cases such as lighting projects, the cost of this "performance insurance" is probably not justified.

Second, ongoing performance guarantees will require the Utility to make certain guarantees to the financiers which may not be possible under the Utility's regulatory structure or may require regulatory approval which could significantly delay the project or discourage the Utility from entering the Agreement in the first place.

Government facilities are encouraged to consider whether the additional cost and time involved with an ongoing performance guarantee (beyond the Utility's performance warranties at Acceptance and possible ongoing O&M contract guarantees) are worth the additional cost. If the answer is yes for all or part of the project, the Contracting Officer should raise this issue as early as possible in the negotiation process.

WR.3 Utility Limitation of Liability—The limitation of liability language found in this Section is standard language for any construction contract. It stipulates that the Utility will be responsible for insuring the ECM operates as designed but that the Utility will not be responsible for indirect or consequential damages that may arise if the ECM operates improperly. In addition, this Section points out that the Utility will not be liable if damages arise from the Government's negligence.

WR.4 Utility Default—Self Explanatory

WR.5 Prompt Payment—Self Explanatory

WR.6 Disputes—Self Explanatory

WR.7 Differing Site Conditions—Sections WR.4-7 paraphrase and reference FAR provisions that are often included by reference in contracts between Utilities and the Government. They are explicitly cited in the Civilian Model Agreement because they are important issues to the parties as well as financiers.

FINANCING AND PAYMENT PROVISIONS

FP.1 Energy Savings and Financing—It is envisioned that the savings from the ECM will exceed the payments to the Utility. The Civilian Model Agreement recognizes that the repayment period for financing of the ECM should not exceed 10 years.

FP.2 Financial Incentives, Rebates, and Design Assistance—This Section confirms that the Government will receive the same rebates, incentives and other services that other Utility customers in the same class receive.

FP.3 Calculation of Payment—Government payments will be determined using the ECM Cost. These payments will not begin until two events have occurred: 1) the Government takes Possession of the ECM and 2) ECM Performance Verification Testing is successfully competed.

FP.4 Buydown—It is critical to the financiers that the consequences of a buydown be clearly defined in the agreement between the Government and the Utility.

FP.5 Pre-Acceptance Termination—As noted elsewhere in this Explanation, financiers want to know with certainty what they will be paid should the Government terminate the Task Order prior to Acceptance. It is impossible to develop a schedule that would be able to predict how many construction dollars would have been spent at any one point in time. Therefore the Civilian Model Agreement recommends that the parties develop a formula that can be applied to the actual dollars spent to determine what the Government will pay should it decide to terminate the Task Order prior to Acceptance.

FP.6 Post-Acceptance Termination—This Section describes what will happen should the Government terminate the Task Order. It recommends that a Termination Schedule be developed by the parties so that there will be a pre-agreed sum the Government will pay upon termination. The creation of a Termination Schedule will be critical to attracting financing and will be required by the financiers.

FP.7 Assignment of Claims—In most cases, Utilities will finance ECMs using third party financiers rather than internal cash. The financiers provide the money for the project in exchange for the Utility assigning to the financier all of the payments the Government is obligated to make under its agreement with the Utility. In order for an assignment of claims to be successful, the parties need to use specific language and follow specific procedures.

FP.8 Novation—A novation agreement is one in which the parties agree that the Civilian Model Agreement shall be binding even if the Utility is purchased by or merges with another company. The Civilian Model Agreement recommends a novation clause be considered, but it is not necessarily required.

SPECIAL REQUIREMENTS

SR.1 Environmental Protection—This Section and the ones that follow set out the Utility's and Government's obligation regarding environmental protection issues and hazardous wastes.

SR.2 Environmental Permits—It is the Utility's responsibility to acquire all of the necessary environmental permits and the Government's responsibility to help where necessary.

SR.3 Handling and Disposal of Hazardous Materials—It is the Government's responsibility to handle and dispose of pre-existing hazardous materials. The Government agrees that the Utility has assumed no obligation to search for pre-existing hazardous materials and if the Utility discovers such materials during its work, it shall stop Work and notify the Government.

SR.4 Asbestos and Lead-Based Paint—Asbestos and Lead-Based Paint may be the exception to the rule that the Government is responsible for removal of hazardous wastes. This Section offers some suggestions as to how the Government and Utility may agree to have the Utility perform the testing, removal or abatement of lead-based paint or asbestos.

SR.5 Refrigerants, Fluorescent Tubes and Ballasts—Self Explanatory

Index